Return of the Gar

Return of the Gar

Mark Spitzer

Number 3 in the Southwestern Nature Writing Series

Denton, Texas

Printed in the United States of America.

10 9 8 7 6 5 4 3 2 1

Permissions:

University of North Texas Press

1155 Union Circle #311336

Denton, TX 76203-5017

The paper used in this book meets the minimum requirements of the American National Standard for Permanence of Paper for Printed Library Materials, z39.48.1984. Binding materials have been chosen for durability.

Library of Congress Cataloging-in-Publication Data

Spitzer, Mark, 1965- author.

Return of the gar / Mark Spitzer. -- Edition:first.

pages cm -- (Number 3 in the Southwest nature writing series)

A sequel to the author's Seasons of the gar
(Fayetteville: University of Arkansas Press, 2010).

Autobiographical.

Includes bibliographical references.

ISBN 978-1-57441-599-5 (cloth : alk. paper)

1. Gars. 2. Gars--Conservation. 3. Alligator gars--Southern States--Conservation. 4. Spitzer, Mark, 1965- I. Spitzer, Mark, 1965- Seasons of the gar. II. Title. III. Series: Southwestern nature writing series ; no. 3.

QL638.L4S64 2015

597'.66--dc23

2014038547

978-1-57441-607-7 (ebook)

Return of the Gar is Number 3 in the Southwestern Nature Writing Series

The electronic edition of this book was made possible by the support of the Vick Family Foundation.

For those dedicated to the science, fishery management, research, and sustainability of all our past, present and future gars.

Table of Contents

List of Photos

Return of the Gar

Introduction

The Gar Returns

The number one question I get asked about gar is why I'm so interested in them. My primary answer is that they're the coolest fish I've ever seen. I mean, just look at these things: They're from a tubular, fossil-fish family that's been around for over a hundred million years, they have an arsenal of deadly fangs, they have armored scales, and they can breathe air with lung-like organs. My secondary response, however, has to do with the mythology of this fish, which has historically been labeled a monster. Having always been intrigued by our fascination with creatures we attribute "supernatural" qualities to, I couldn't help actively investigating this symbolically rich, dragon-headed fish.

Along the way, I studied the science, the history, and the folklore of gar. I fished for them, wrote and published "garticles," and explored their eco-issues. My research got picked up by the fishing celebrity Jeremy Wade of *River Monsters* fame, I caught gator gar with him and appeared on his show, and I also consulted for the Zeb Hogan *Monster Fish* episode on alligator gar produced by National Geographic. Around the same time, I hooked up with the international gar community, comprised mostly of biologists and fishery specialists. Then, in 2010, my first gar book, *Season of the Gar: Adventures in Pursuit of America's Most Misunderstood Fish,* came out from the University of Arkansas Press.

That book deserves mention, since it's technically the prequel to this one. *Season of the Gar* took a look at why gar have been so vilified in North America, and it proposed some ideas for preserving and propagating the species. It also included some semi-Gonzo first-person narratives about fishing for gar with a goofy character named Hippy. Essentially, that project was the first book-length study of gar in the English language, accessible to both a specialized and general-reading audience. It provided a lot of information in order to put gar into perspective—which is what I'm aiming to do with this intro, but in a condensed way, so that we can get on with a new discussion regarding gar.

But first, there's another frequently asked question about this fish: How big do they get?

The answer to this is that alligator gar are the second largest freshwater fish in North America, and we have reports of them reaching lengths of eleven feet, maybe more. Nine-footers have been caught, and some have been documented weighing well over three hundred pounds. The Cuban gar is the next biggest species, which can grow to be two meters long. Longnose gar can also get pretty big. I've gillnetted six-footers weighing more than thirty pounds, but there are some on record weighing over fifty. Tropical gar can supposedly grow to four and a half feet long. The spotted and shortnose come next, maxing out at about three feet and seven pounds. Typically, the Florida gar is the smallest gar, though they can sometimes surpass spotted and shortnose in length and mass.

The seven gar species mentioned above are members of the *Lepisosteidae* family, which basically translates as "bony scaled fish." There are other fish in the world called "gar" (i.e., Atlantic gar or needlefish, marbled or Peruvian gar, etc.), but those are entirely different species. The gar I'm talking about have hard, diamond-shaped, ganoid scales, and they're endemic to North and Central America and parts of the Caribbean.

In the 1800s, the settlers in what's now the United States were pretty freaked out by the alligator gar. Their food value was dismissed, and because of their scary-looking faces and man-eating size, they were quickly demonized as out-of-grace "devil fish." Scandalous stories arose as the God-fearing masses ascribed qualities of Satan to a harmless, passive species. Due to this stigma, gar were essentially cast out to make way for irrigation and agriculture—spawning grounds be dammed! A species that once ranged across the continent and even into Canada was then relegated to the South, where warmer waters and swamps provide spawning grounds. Sure, some populations of alligator gar still exist above the Bible Belt, but for the most part, gator gar were extirpated from half of their natural range in a few hundred years.

In Arkansas, a century of general hatred for these species combined with industry and development came to a head when "sport fishing" was thrown into the mix. Like many populations from Ohio to Texas and everywhere in between, gar had already been shot up, blown up, and burned en masse by the mid-twentieth century. So when Arkansas started advertising spectacular big game fish to be caught on heavy-duty tackle, it only took three years in the late fifties for the top freshwater predator in the state to get wiped out. Because of this, the ecosystem suffered. With no gator gar to thin the habitat-decimating fish populations (buffalo, carp, drum, etc.) that other fish couldn't swallow, the big members of the minnow family grew even bigger, destroying the eggs and nests of other fish.

In the last two decades, though, gar science has mushroomed exponentially, thereby replacing faulty and biased science with constructive information. We found out that gar eating twice their weight in game fish per day was simply not true, as well as rumors that they attack humans.

But the damage had already been done, thanks to states like Louisiana, Texas, and Illinois encouraging citizens to destroy gar in the thirties. Such propaganda led to a culture of smearing gar. Fathers taught their sons to never return gar to the water, and generations of breaking off beaks and leaving lunkers lying on the shore persevered for decades.

In a nutshell, that's what the prequel was about. This book, however, is more optimistic. Now that awareness is on the rise, this book is concerned with bringing crippled populations back up to snuff. We've been doing this in Arkansas, as well as Missouri, Texas, and throughout the South, thanks to a coordinated effort between federal and state agencies, universities, hatcheries, independent researchers, and the media—and the results are visible and tangible.

But what about gar outside our borders? That's another question this book asks as I travel to Central America, Thailand, and Mexico to assess the global gar situation.

Still, it would be misleading to suggest that this is mainly a fish book, because the questions I'm asking about gar (spoiler alert!) are metaphorical for the questions I'm asking about our own species and our planet during a time of crisis. Global warming is real. Our ice caps are melting at the rate of one percent per year, "superstorms" and droughts are increasing, flooding is a lot more common, and sea levels are on the rise.

But like I said, this book is optimistic. It evaluates what we have done for gar, what we can still do for gar, and through association, what we can do for ourselves and future generations—because gar aren't just fish. They're also a reflection of our struggles with each other, which reveal themes of oppression, violence, ignorance, and ultimately, stupidity. Because wouldn't it be stupid to continue to regard this part of our natural history (so therefore ourselves) as disposable enough to fritter away?

Of course. But that's what we're doing every day as if we're not all bound by an intricate network of organisms we casually call "the ecosystem." As if the word "eco" doesn't mean *home*. As if our home can be as colorful and vibrant as it used to be if species diversity becomes an aspect of the past.

I know I speak for millions when I write that we don't want to live in a world of dwindling fish. We want a world full of healthy creatures, one that took eons to evolve. We want a world that has a place for the imagination, education, ingenuity, co-existence and civility. And we can have that world, where we keep all that, and foster all that, if we're smart—which is what we're finally being in relation to alligator gar. They need us, we're helping them out, and they're beginning to return.

But who really cares if something comes back? Unless, of course, that something is useful, or has a message to apply—which is what this book is all about: looking at what's working and doing more of that, and looking at what's harmful and doing less of that. And if readers end up with more questions than they started with, that's okay. It's the process of trying to understand this fish that will ultimately lead to a less threatened status for the most endangered gar species: alligator, Cuban, and tropical gar.

* * *

I'm no scientist. I'm a writer from the arts and humanities who has a respect for science, history, folklore, and politics. I note this not as a disclaimer, but to make clear what my intentions are with this book. It's a general overview of what has happened, what's happening, and what can happen with gar, and its audience is made up of vast and diverse demographics. I expect that anglers and people with strong connections to the outdoors will be the readers most interested in the following pages. I also expect that students and teachers and avid readers of creative nonfiction and ecology books will make up another good percentage. With this in mind, I've tailored the following narratives for a general-reading audience, though I'm sure that there are those in specialized fields who will find this work useful.

Biologists tend to be my most critical reviewers, but I don't write in that lexicon. I write from the stance of a creative writer with a background in storytelling. That's how these chapters are constructed: through narratives that utilize research, images, facts and mythologies.

With books like Mark Kurlansky's *Cod: A Biography of the Fish that Changed the World* (Penguin, 1998) and Paul Greenberg's *Four Fish: The Future of the Last Wild Food* (Penguin, 2010) as models that focus on our relationship with fish, my objective is to cover as much ground as possible through a mode of discourse that, stylistically, verges on old school New Journalism—a form of nonfiction that takes risks, plays with language, and doesn't shy away from embracing bias. But to be a reliable narrator in this day and age, one must be extremely cautious with one's opinions. Hence, I've tried to picture my personal views as emotional responses that happened to happen, rather than the way it is. Meaning this is not a textbook. If anything, it's a highly researched vision that is meant to entertain and inform through honest, action-packed narratives that relate experiences in local and global environments.

That said, I'd like to thank the University of Central Arkansas for a University Research Council grant in support of travel for this book, and for a sabbatical for finishing the manuscript. I'm also grateful to Daiichi Hooks and Tackle and Penn Rod and Reels for sponsorship in the form of fishing gear—a connection made possible by my good friend Keith "Catfish" Sutton. Members of my "Fishing Support Group" also deserve acknowledgement for going after gar with me: Ben "Minnow Bucket" Damgaard, Tim "T-Bone" Thornes, Tyrone "Ty-Stick" Jaeger, Rob "Turkey Buzzard" Mauldin, and occasional special guest star Scotty Lewis. As for the editing and production of this book, I'm glad to have had the highly professional assistance of editors Karen DeVinney and David Taylor, and manuscript readers Dr. Susan A. Cohen and Dr. Solomon David for the University of North Texas Press, a publisher I'm proud to work with.

Gar experts Lindsey Lewis of US Fish & Wildlife, Dr. Alysse Ferrara of Nicholls State University, Dr. Reid Adams at the University of Central Arkansas, David Buckmeier of Texas Parks & Wildlife, Dr. Roberto Mendoza at the Universidad Autónoma de Nuevo León in Mexico, and biologist Ed Kluender were extremely helpful whenever I had questions about gar. I've already mentioned Solomon David at the Shedd Aquarium in Chicago, but I'd like to mention him again since I consulted him constantly during the revision process. I was also fortunate to have had access to first-hand information from US Fish & Wildlife biologist Tommy Inebnit. Fishing guides and gar authorities like Jim Dussias, Jules Fernandez, and Mino and Nook also played a key role in helping me land gar. Hector Garcia was invaluable, writing letters, translating Spanish, and coordinating with Cuban and Mexican contacts. Prof. Gabriel "Gabo" Márquez Couturier in Tabasco, Mexico took me on a whirlwind tour of a half-dozen gar farms, took me out for *pejelagarto*, and provided an insider's perspective on tropical gar few gringos have ever been privileged with. Thanks also to Jennifer L. Bouldin, Associate Professor of Environmental Biology at Arkansas State University, for responding to questions about toxin levels in Lake Conway, along with my undisclosed contact connected to the Arkansas Department of Environmental Quality for inside information on the Mayflower Oil Spill report.

Versions of chapters in this book have been published by *The Oklahoma Review, Terrain, Animal, Flyway, The Louisiana Review, Green Hills Literary Lantern,* and *Big Muddy,* and I thank those editors for their insights.

Ultimately, though, I'm glad to thank the fish that inspired this book. Sure, it's a ludicrous thing to do—to thank something that can't thank back—but maybe that's not really the case. Maybe gar have their own way of being grateful, and we just don't have the capacity yet to recognize such responses. I'm hopeful that's the way it is.

—Mark Spitzer, 2014

Chapter 1

The Spawn and Beyond

A Metaphor for Sustaining Biodiversity as the *Deepwater Horizon* Spews into the Sea

When I pulled up to the flooded field, there was already a line of pickups parked on the edge of it. I could see a bunch of people standing in the shin-deep water where the river had consumed the gravel road. Lindsey Lewis, gar specialist from US Fish & Wildlife, was out there with three bowfishermen, and I could also see Ed Kluender, a graduate student in biology at my university who specializes in tracking movement patterns of gator gar in Arkansas. Ed had called me the night before and told me that the spawn was on. I told him I'd meet him at 8:00 a.m., so that's why I was there—with my canoe.

I passed a couple of the bowhunters' wives, also standing in the water, and waded out to Lindsey and Ed. They were focused on the ditch alongside the upstream side of the road, as were the bowhunters, whose bows were lowered. But before I could even say a word, a mammoth back breached like a submarine. I could see its spotted tail and dorsal fin, and enough of that greenish-black checkered pattern on its gun-metal-gray armor to know it was a six-footer, at least.

Then it shot out onto the road, zigging and zagging toward the trucks through five or six inches of water. It was shooting straight for one of the women. Ed was chasing after it, so I lit off after it too.

The woman screamed and turned tail, but the gator gar kept coming at a clip faster than she could run. That gar was slashing and splashing and whipping its tail like a crocodile with road rage. Basically, a 200-pound human was trying to escape getting run over by a 150-pound prehistoric monstrosity, which was looking unlikely at the moment. So resigning herself to her fate, the fishwife froze in her tracks and braced for impact. The gar shot right between her legs, then beached itself on the road, before twisting at the last second and torpedoing downstream.

Ed was out of range, so it was up to me to try to slow it down, then hold it till the others could assist. I didn't know why they wanted to catch it, but I figured it was to tag it and attach a transmitter, or check to see if it had any tags—because every speck of data we can get is key in protecting and preserving the species as a whole.

I'd been catching and releasing alligator gar with "Team Gar" (that's what Lindsey called us) for the last two years. The process usually involved spreading gillnets in the winter when the frigid currents made the fish a lot more docile and easy to handle. But this time was different. This time I was hellbent on tackling a monsterfish.

Then, right when I leapt, the road disappeared beneath my feet. Having run straight into the ditch, which I couldn't see because of the floodwaters, I stumbled, fell, and the gar made its getaway.

After that, it was a lot of hooting and knee-slapping as we recounted the events we'd just seen—and those we hadn't. Apparently, four mongo gar had just spawned out in the field. There was a small creek going through a culvert under the road we were standing on, which the gar had swum through the night before. The water was going down, and the gar knew they had to make it back to their home stream, a tributary of the Arkansas River. They also knew that humans were there, complicating the process.

I looked around. On the other side of the underwater culvert, where the gravel of the road emerged, I could see three fat buffalo fish that the bowhunters had left for the flies—as they had intended to do with that gator gar that got away. No doubt, they'd been waiting here since before the sun rose.

The bowhunters left, and Ed told me how he'd been bummed to see them standing there with bows raised when he pulled up, about forty minutes before I arrived. They were waiting for the big ones, as they'd done for years whenever the river rose in the spring. Ed, however, was even more bummed to see them suddenly shoot into the ditch.

A thrashing ensued, Ed ran out, and because he was closer than the bowhunters, he got to the gar as it was crossing the road. Ed jumped on that gar, got right on its back, and yanked the arrow out of its head. As he did this, he told the bowhunters that he worked with US Fish & Wildlife, and that it was illegal to shoot alligator gar during their spawning season. No apologies whatsoever!

Inspecting the fish, Ed let it go. It was an ancient seven-footer. The bowhunters had injured it, but Ed figured it would make it. He called Lindsey on his cell.

Lindsey was literally out in the field, half a mile upstream in a kayak. He'd been videotaping the spawn, unaware of the bowhunters, who were unaware of the new rules.

The summer before, the Arkansas Game and Fish Commission had passed a package of new harvesting laws. These days, a permit was required; there was a new one-gator-gar-per-day limit; and all alligator gar over three feet long were off limits during May and June. Ed had explained this to the bowhunters, who were pretty much shaking in their waders when Lindsey had pulled up with his uniform on.

Lindsey, representing the Feds, then scared the crap out of those ol' boys by explaining how much trouble they were in. He could've brought the hammer down on them, but instead he let them go with a warning and told them to tell their bowshooting buddies about the new rules. That's when I happened upon the scene.

The main thing, though, was that our gar had spawned, which doesn't happen every year. When it does, it's due to a harmonic convergence of the right water level and water temperature, and the fact that those fields have to stay flooded for at least ten days for the fry to make it on their own. In this traditional spawning ground, where gar have been breeding for centuries, successful spawns tend to coincide with major flood pulses, which don't happen every year. Dr. Reid Adams at the University of Central Arkansas told me in an email that for this particular population "analysis of historical hydro data suggested that hydrological regimes were conducive to 'good' reproductive success, including recruitment, of alligator gar around 5 years out of 35." This data comes from Tommy Inebnit's 2009 thesis on this population, "Aspects of Reproductive and Juvenile Ecology of Alligator Gar," and Reid was his advisor. My overall point being: on average, this population of gator gar only has a successful spawn once every seven years, and when this happens, a very small percentage of the hundreds of thousands of eggs that get jettisoned survive to become breeding adults.

For an hour or so, we waited for more gar, but only saw spotted and shortnose crossing the road. The smaller species were also heading back to their home flow. Just for fun, Lindsey threw a cast net out, but failed to capture any of them.

Lindsey figured the big ones were gone, and that the one that almost ran down the woman was the last of the four he'd seen spawning.

"It was incredible!" Lindsey explained, animated by adrenaline. "There was one female and three males, all of them at least six feet. I got right next to them and they didn't stop. Just kept splashing like crazy, sending plumes of water ten feet into the air. Even the cows were watching. And a deer."

This made me think of how monster myths are spawned as well. From Lake Champlain on the border of Quebec circa the early 1600s to Lake Norman in Oklahoma to almost everywhere there are alligator gar, the ruckus they kick up when they mate is enough to make anyone think the disturbance is coming from an unknown leviathan. These stories go through a few narrators, the details get exaggerated, newspapers pump up the hype, and in the end we get tales of raging, man-eating creatures—when all it was was a fish orgy.

Anyway, that's my theory and I'm sticking to it.

* * *

Ed and I followed the trees in my canoe, tracing where the creek used to be. I was paddling in back and he was up front, and between us we had an empty five-gallon bucket. When the tree-line forked, we veered toward the area where Lindsey said the gator gar had spawned.

We could see the flattened tall grass, where the fish had rolled. It was about two feet deep. Then we saw the eggs.

They were a translucent orange, the size of BBs, and scattered in a trail through the thrashed grass. We followed their path to where the vegetation was most disturbed, and came upon the mother lode. At least 157,000 eggs were stuck to submerged plants—that's the number gar researcher Dr. Alysse Ferrara, in her 2001 dissertation "Life-History Strategy of Lepisosteidae," indicates are released from a typical female.

I asked Ed what kind of grasses these were, and he said "some sort of sedge called *Cyperus*." They were about three feet tall and had spindly heads.

We got to work, collecting as many stalks as possible, covered with the sticky eggs. We began filling the bucket with hundreds of clusters, since we knew what was going to happen: the field would dry out in a day or two, and the entire spawn would be totally shot.

Gar, however, are our people, so we weren't about to let that happen. Once we got that bucket full, we paddled back to the underwater creek, found some brush sticking up, and stuffed the eggs as far down as we could reach.

If the water went down, this was our assurance that a few would have a chance. That is, if the egg-sucking scourers patrolling the floodplain didn't discover them first. Having fished this system obsessively, I knew it was full of buffalo, drum, and carp, which are the major prey of inland alligator gar—and the real culprits behind gamefish nesting devastation.

Realistically, we knew that the odds were pretty low that these eggs would hatch, since these fields were also filled with snakes, turtles, crawfish, and hundreds of other creatures that might be able to devour them. At least we could be confident that it was highly unlikely that any mammals or birds would gobble them down. If they did, then they'd suffer the fate of many a chicken, dog, and human who found out too late that gar roe is extremely toxic.

The Aaron family of Cleburne County discovered this the summer before when they went spearfishing in Greers Ferry Lake and bagged a bulging longnose and decided to make caviar. After their meal, it only took a few hours for the poison to kick in. It started with vomiting, then got worse. They were rushed to the emergency room, where they shook and sweated and almost died, but eventually pulled through. This is a story I hear every spring.

It's a good thing that alligator gar have this built-in defense mechanism, since their reproductive process is extremely complicated. A few years ago I interviewed biologist William G. Layher (author of the 2008 study "Literature Survey, Status in States of Historic Occurrence, and Field Investigations into the Life History of Alligator Gar in the Ouachita River, Arkansas"), who told me that if spawning conditions aren't optimum, the eggs might never get released. I asked where they went and he said they are sometimes absorbed back into a fish's system. Hence, it pays to produce a poison so fierce that those with the munchies stay away from eggs that have a challenging time making it to maturity.

In this sense, gars have the capacity to be lethal. There have been no documented cases of an alligator gar ever actually attacking a human (just questionable stories), but there are plenty of verified instances of people being hospitalized for chowing down on deadly roe.

Nevertheless, the big ones have earned a reputation for being razor-fanged sociopaths viciously seeking humans for brunch. This stereotype, however, is simply ignorant. Such rumors were spread by early American gar-fearing cultures that demonized a scary-looking fish. This attitude encouraged decades of faulty science and disinformation that continues to endure.

I've made this case many times before, so I won't repeat myself, except to say that luckily, and finally, we're starting to get a fix on gar. The science is picking up, federal and state agencies are working with universities, and there's a serious coordinated effort underway to educate the public on the role of gar in ecosystems—because, as gar specialist Lee Holt of Arkansas Game and Fish once stressed to me in an interview, the more gar you have in a system, the healthier that system is, and the bigger the game fish get. That's a fact, and there are studies to prove it, most notably Dennis L. Scarnecchia's "Reappraisal of Gars and Bowfins in Fishery Management," published in the July-August 1992 issue of *Fisheries*.

There's also been a recent surge of fishing shows on cable and satellite spreading the word that gar aren't harmful. One of these is Zeb Hogan's 2010 *Monster Fish* episode produced by National Geographic, and the other is Jeremy Wade's 2009 "Alligator Gar" episode of Animal Planet's *River Monsters* series. Both of these shows shoot to dispel damaging myths, and both of these shows have informed millions of viewers about misinformation regarding gar.

I was consulted on both of these projects, and my research was used in both productions. Though I have some minor complaints about how the information I provided was used, I'm pretty happy about the changing reputation of the alligator gar. After all, my interest in this subject isn't meant to serve myself; it's to preserve a super-cool, air-breathing fish for future generations, so that it can continue to captivate imaginations and turn kids into engaged and informed citizens who are part conservationist, part something else.

Because that's what we need to sustain the natural world: a persistent interest in preservation coming from a variety of views and backgrounds—especially from those that oppose each other on a vast array of differing issues. Our common ground is our glue, so logically, we should work together on what we can agree on to conserve a major part of ourselves. After all, we are a wilderness nation, founded in the wilderness, carved from the wilderness, our identity formed in the wilderness. To forget this is a travesty which will divest us of this quality, which we can already see diminishing at alarming rates. Just look at the Pacific salmon crash in 2008; the starving, swimming polar bears; the vanishing buffer of marshy grasses on our Gulf shores that protect our cities from hurricanes. Ideally, if more people with differences took more of an interest in our interconnecting network of flora and fauna, we'd have more reason to come together so that we don't come apart.

Again, that's my theory and I'm sticking to it.

* * *

The International Gar Conference was organized by Alysse Ferrara at Nicholls State University in Thibodeaux, Louisiana. There were experts from Mexico, Mississippi, Alabama, Arkansas, the whole gar-dammed world, and I was the keynote speaker.

Season of the Gar: Adventures in Pursuit of America's Most Misunderstood Fish had just come out from the University of Arkansas Press. It was the first book ever published on this subject in the English language, so of course I was worried that my creative nonfiction approach to the species might not fly so well with biologists and fishery experts who refer to gator gar as "*Atractosteus spatula.*" Plus, my chapters on science and folklore and history were interspersed with some semi-ludicrous misadventures.

In a sense, I was putting my book to the scientific test. If these authorities could accept it, then maybe I'd done something good—the Henry David Thoreau quote constantly repeating in my head, "Don't just be good, be good for something."

I was also a bit nervous because I felt I might have left a less than professional impression on some garologists from my state. The evening before the conference began, my wife and I had gone to a pizza place downtown, where we happened upon Lindsey and a bunch of Game and Fish agents eating dinner. I went over and introduced Robin, then made a blundering sweeping gesture in which I knocked over a glass of Sprite, spilling it all over the table. Needless to say, I was embarrassed by that. But they, of course, just laughed it off.

This was one week after the gar spawn in Arkansas, and the book had only been out two weeks. I knew that some of the scientists had already read it, so I was ready to receive criticism. But I was also psyched to take in the panels and talks, hobnob with the world experts, and learn as much as I could.

Anyhow, the next morning the conference began, and suddenly I was up on stage reading my final chapter, which I chose for its science-based optimism about many of the new management plans designed to protect alligator gar, a creature which the American Fisheries Society categorizes as "imperiled" throughout its range (see *Fisheries,* vol. 33, issue 8, 2008). Meanwhile, there was a Spanish translator in the booth to my right, speaking into a microphone. I have no idea how she kept pace as I read in my regular rapid gait, but I could see the Spanish-language listeners nodding with their headphones on.

After that, it was out to the hallway where the thirty books I'd brought along had already been sold out. In fact, there was a line of ichthyologists waiting for me to sign their copies.

The next twenty minutes was one of the most reassuring blurs I have ever experienced. Half of the authorities there had already read my book, which they had pre-ordered on Amazon, and they were full of praise for what I had done, and for how I had presented their body of research in layman's terms, thereby increasing our understanding of the species. And the other half, they were eager to read it, and thank me for spreading their discoveries to those who still bust off their beaks.

A series of presentations followed, but because the list of researchers includes over a hundred names, it wouldn't serve the purpose of this book to list them all in this chapter. Still, the cutting-edge work of these scientists and fishery experts should definitely be recognized, especially in any timely book-length study regarding gar. What's important at this point is to show the highly sophisticated work going in the world of gar, so I'll mention just a few names here, and include the entire conference schedule in the "Garpendix" at the end of this book.

The first panel concerned the restoration of tropical gar in Costa Rica. Then came a panel on gator gar management activities in Alabama and a PowerPoint on pre- and post-regulation harvest rates on the Trinity River in Texas. New and stricter laws had just been passed in that state, and officials were looking at how tournament bowhunters were working with Texas Parks & Wildlife to provide specimens for research now that the commercial fishing industry for alligator gar had been severely reduced, thanks to the new daily limit of one per day.

Then Ed presented his highly awaited talk, because the gar research at the University of Central Arkansas (where I ironically scored a job) is pretty much considered the avant-garde in gar studies. Ed's presentation was entitled "Movements and Habitat Use of Adult Alligator Gar in a Tributary of the Arkansas River," and I was impressed with how he articulated his findings to those whose degree he was seeking.

One downer, though, had to do with the spawn we'd just witnessed. Ed was showing pictures of the field we'd been in last week, plus graphs associated with weather and water temperatures, and then he got to the water level. Like everyone around me, I cringed when he told us that the spawning ground had dried up two days later.

This, however, was nothing compared to the big unfortunate news of the day, and the big unfortunate news of the summer. For the last month, the out-of-control BP oil spill had been filling the Gulf of Mexico with 53,000 barrels of unrefined petroleum per day. And as everyone in that auditorium knew, alligator gar live in salt, fresh, and brackish water, and their main populations line those shores. In fact, Gregory P. Moyer and Brian R. Kreiser projected a map of North America from their "Preliminary Analysis of Range-Wide Population Structure in Alligator Gar" study, with the main populations marked by red flags clustered all over the coast. And as everyone in that auditorium also knew, we were now in the middle of spawning season, and it didn't look like the leak could be stopped. The prediction was that the *Deepwater Horizon* well would spew for the entire summer, it would poison and pollute the Gulf, it would gum up the beaches and vegetation, and it would kill off the crustaceans (like blue crabs) that the saltier salinity gar rely on as the main staple of their diet. And that's exactly what happened.

We even had a special panel in which a US Fish & Wildlife official in the midst of this battle stopped to talk to us about how this spill could affect gator gar specifically, and what we can do about this. Whatever the take-home message was, it escapes me now, but I know that a lot of us were extremely concerned about what this meant for the fate of our finny friends.

But then, at least, some good news came in. I was out in the hallway between panels when somebody thrust a cell phone in my face. Reid Adams, the project leader on all things gar at UCA, was on the line. He hadn't been able to make it to the conference, and he was kayaking up the creek where Ed and I had stashed the eggs in the brush. He wanted to know exactly where we'd stuffed them.

It took a few minutes to guide him there, and when he got to the spot—

"There they are!" he exclaimed. "They made it!"

He could see the little black fry swimming near the surface, with their tiny gold racing stripes running down their backs.

I immediately told Ed, who immediately got up on stage at the next panel and told the gar-crowd what Reid had discovered. A cheer erupted and we all felt some sense of hope.

And so the conference continued, with talks on hormones in breeding and their effect on gar larvae, grading frequencies on alligator gar fingerlings, and characterizations of the supply network of tropical gar in the region of Tabasco. Optimal feed rates for juveniles reared in recirculating systems was covered along with the results of rearing gator gar spawn using different culture systems in Costa Rica. We then heard about the aging and growth of Florida gar, countergradiant variations in growth of spotted gar from different latitudes, and the effects of ambient salinity on plasma osmolality. Effects of salinity acclimation on growth was investigated and so were mercury concentrations in muscle tissue of longnose in North Carolina. A panel on teleost genome duplication for spotted gar as well as functional analysis of *sox9* (whatever that is) in spotteds (or "spots" as they're commonly called in the field) presented their findings, followed by a discussion of digestive physiology to design micro-diets for gar. There were also talks on temperature studies on alligator gar and a presentation on physiological responses of gator gar to pollution.

The next day it was all about bacteriocidal activity of spotted gar serum, reproductive characterization of spotted gar in Louisiana, evaluating habitat utilization and diet, analysis of diets in drainage-canal gar, neurotoxic potency of gar oocyte extract peaks at spawning, strategies for the commercial pilot scale culture, small-scale experimental culture and cost analysis, PVC-lined circular tanks used for farming gar in Mexico, some international gar business meetings, an aging workshop, a poster session, and everything a gar-nerd could ever dream of.

All this culminated with a crawfish boil in which Ricky Verret—a commercial fisherman from Houma who specializes in catching alligator gars on jugs—was awarded a special prize for supplying specimens for study and collecting data. Ricky then led us in "the gar dance," which was a line dance in which you flay and chop and skin an invisible gar while a Zydeco band jams into the night. And as game and fish agents from across the continent played some sort of game with a tree stump and a hatchet (in which my small blonde wife showed those burly dudes a thing or two), the last keg was eventually drained.

* * *

All the way back to Arkansas, we listened to the disaster reports. The dolphins and pelicans were washing up with the tarballs, the shellfish were ingesting the gunk, the shrimp and oyster industries were taking a huge hit, fishing was halted throughout the Gulf, and now toxic dispersants were being used to break up millions of gallons of cruddy crude with no introspection whatsoever as to what the short- and long-term effects of the undisclosed chemicals could be.

But at least I was heartened by the fact that during the worst man-made environmental disaster this country had ever seen, a united coalition was working on gar. It was a collaboration of state and federal authorities, university scientists, independent researchers, hatchery and fishery experts, commercial and recreational fishermen, students, scholars, fish farmers, engineers, the works. And with public education programs in Missouri working as a model for programs in the southern states, we were reaching out to school-age children, who influence parents, communities, and the future of our fisheries.

In fact, more changes for gar in favor of gar had been made in the last ten years than in any time in history. More laws and programs are now in place, awareness has increased, and the serious science is finally being funded. Because that's what happens when a species starts going down.

Same thing with global warming. Ten years ago the idea was debated, but now the results are in and the only naysayers debating this issue are those in denial, or under the thumb of the petrochemical industry. So as the water rises and the jet streams swing erratically, instead of bickering constantly, we find ourselves searching for solutions—because we have to.

Teaching environmental literature and eco-writing for the last decade, I've been watching our actions closely. We've gone from yowling "The Sky Is Falling!" to trying to learn how to live with our declining condition. In a way, this plight parallels that of modern cancer, which (to generalize) is now less of an imminent death knell and more of something to treat as long as possible.

This statement is no doubt controversial, and I'm sure there are those who will object, but like almost everyone on this planet, I have been affected by disease. We all know people who've had cancer, or have it now, or will have it in the future. According to an article by Zosia Chustecka on the Medscape website (February 9, 2007), one in three women gets cancer these days, and for men, the probability is 50 percent. At this point, I'm not so sure my mother, who has terminal cancer, will ever see these words in print.

Thus, the trick is to live with it: cancer, disease, the changing climate. But the trick is also to look ahead, beyond the spawn, to the next one, and those that follow as species die out, as development continues, as our needs change and those of gar stay the same.

The spawning grounds are diminishing at an exponential rate. The more dams we have, the less flooding there is. Still, as Tim Radford notes on the Climate News Network website (June 10, 2013), we're experiencing more once-in-a-hundred-year-floods now than we've ever had before, since they're coming at the rate of once every decade. The upshot being:

our system is confused. So we need to think long-term as we factor this confusion in, which is a whole new way of looking at the problem.

How we do that, I don't know—but it's something to start thinking about way more creatively than we've ever done in the past. I do know, though, that Lindsey is on the right track. He's working with the farmers who own the fields. He's explained to them what we're doing, what we have to lose, and what we have to gain. And they, in turn, have granted us access to their land.

Lindsey is also working with the Corps of Engineers who run the dam that controls the level of the tributary we primarily work on. Plus, he's put in for some grants to install custom culverts under the road where those old guard gar tried to get past the bowhunters. Such culverts, however, aren't meant to protect fish from predators; they're meant to provide easy access in and out of the spawning grounds, so that when gator gar are spent and the water level is going down, they don't have to fight the gravel as well.

Question: Why did the gar cross the road?

Answer: To carry on its chromosomes.

And ours.

"Green engineering" might be a partial answer. Out West, dams are constructed with stairways for salmon. In the East, ladders for American eels have been in use for years. So why not gar-ladders? And why not create more protected areas where gar have spawned for centuries, so they can continue to spawn for centuries?

Well, the answer is there's not a will. And where there's not a will, there's not a way.

But I shouldn't say there's not a will; there's only not a *popular* will. Because there is a will behind the scenes. Team Gar represents a will. The entire international gar community represents a will. Down in Pine Bluff, William G. Layher's vision for engineering a spawning ground near the Felsenthal Dam on the Ouachita River represents a will. And in Little Rock, Arkansas Game and Fish Commission agent Lee Holt has a vision for turning an oxbow lake into a marshy gar nursery. That's a will as well—just like Christopher Kennedy's will in Missouri, where he was successful in getting the Department of Conservation to create a gator gar refuge in Mingo Swamp as part of an ambitious alligator gar restoration project.

Photo 1. Christopher Kennedy Releasing Young-of-Year Gator Gar in Mingo Swamp, MO.

Photo Courtesy of the Missouri Department of Conservation.

But it's not just the species that needs a bump. It's the reason behind why gar need a bump which is why we're so concerned with this fish.

Basically, gar make ecosystems stronger, more diverse, more biologically colorful. Meaning gar help us sustain our natural heritage, without which, we'd be losers—of biodiversity and the benefits that accompany that.

More species equals more possibilities in medicine, of course, but that's not what I'm talking about. What I'm talking about is co-existing with the richest type of quality of life possible, in which a kid can catch a salamander and show it off to his friends. In which a father can show his daughter how to reel in bluegills, then clean them, and she can tell the story of catching them at dinner. I'm talking about preserving our natural world as part of our history and our future, because we respect its mysteries.

But as I write this, during the summer of 2010, the oil is spewing into the Gulf, the ice caps are melting at the rate of 10 percent every decade, and as author Richard Ellis notes in *On Thin Ice* (Vintage, 2010), the polar ice caps will be gone in a century. Whether or not this can be proven, various models estimate that the planetary air conditioners (a.k.a. the North and South poles) won't have enough Freon left in a hundred years for the changing jet streams to cool the world.

So that's where we're at, that's how we are, and this makes me feel even bleaker. My faith has flown the coop.

* * *

But then I'm back in Arkansas, paddling up that little creek, which is so shallow that if it doesn't rain in the next few days, everything in it is going down. And then I see it: a one-inch-long, black and gold gator gar! It's stalking some even smaller minnows along the shore where the water is tepid and still. It looks like a tiny stick, barely visible, and it's hardly concerned with me at all.

So I dip in my butterfly net, extend the hoop under it, scoop it up, and let out a "WHOOP!" Because catching this mini-gar is just as much of a blast as hauling in a full-grown fatty.

Then I catch two more, so take them back to my tank, not so convinced we're all going to die. I'm also not so convinced that the idea of gator gar making a comeback is an idealistic fantasy. Because in Arkansas, where alligator gar were essentially wiped out in the fifties, we have made a difference. After five decades of regret and one decade of seriously getting down to business, those three little suckers are living proof that we can bring them back from the brink, and save them—and ourselves —if our will and integrity is as strong as our desire to capitalize on the resources we literally have at our disposal.

And that ain't just my theory, folks. That's the way it really is.

Photo 2. The Battle Begins.

Photo by Mark Spitzer.

Chapter 2

Gar vs. Sewage

A Tragedy of Waste

> "Now, gods, stand up for bastards!"
>
> —*Lear,* I. ii.

It started with a flimsy cardboard sign, impaled in a frosty, windblown field. The sign said that the local utility company, Conway Corporation, had applied for a permit for construction in the area known as Lollie Bottoms. There were no notices in the paper or on the news; just that lonely sign in that stark field informing 52,000-plus citizens that a public meeting had been scheduled to discuss any questions or concerns.

I was alerted to this sign by a home owner who lived along the Arkansas River, a mile beneath the Toad Suck Dam. She had read about my water quality work with the Lake Conway Advisory Group in the local paper, and for some strange recurring reason, she singled me out as an eco-professor who knew how to assume a leadership role when developers rumble in with bulldozers—which was a far cry from the truth.

What I did know how to do was create controversy by calling attention to an issue and writing letters to the editor. Whereas she was concerned about her human neighbors, however, I was concerned about the gar.

Lollie Bottoms contains one of the oldest, undeveloped, traditional spawning grounds for alligator gar in the state. The largest-known population of gator gar in Arkansas lives part time in a tributary of Pool 7, where the City of Conway was proposing to build a new seventy-million-dollar sewage plant, and there were other populations of gator gar that lived full time in the river right where the Conway Corp architects proposed installing the outlet pipe. Meaning that this new sewage facility would be sucking up water from the river, treating it with chemicals, then spewing it right back into the most densely populated alligator gar population in the region.

What concerned me even more were the seven hundred days slated for construction that would muddy up Tupelo Bayou, a creek that flows straight down to the spawning grounds and empties into the Arkansas. The summer before, at least three fully grown gator gars had been shot by bowhunters in that area. In one case, three radio transmitters had been removed and thrown in a ditch—so I knew the big ones were mating in there.

I also knew that gar are extremely sensitive to flow. This was something that Team Gar was in the process of finding out, having sampled from "the Garhole" for the last few winters. When the water was up, we discovered that the big ones moved into the thirty- to forty-foot hole around October, fattened up for the winter, then hunkered down until the spring floods. The previous year, there'd been a drought and the water level in the Garhole had been about thirty feet deep. I'd been out there every month for over a year keeping my eye on gator gar activity, and there'd been a lot less porpoising. In fact, there'd been virtually no activity in that hole for the last year, because the water was just too low.

We also learned that when the water level was low in the tributary and on the main river, the alligator gar didn't congregate in the Garhole. We found this out by spreading nets and getting skunked, but what really provided this information were the different pings biology grad student Ed Kluender recorded throughout the winter of 2010. Basically, the gar we had tagged and hooked up with transmitters were still in the system, but they were spread out about a mile apart, each hunting their own territory.

I was also worried about turbidity. I figured that if the water flowing out of the bayou was mucky and silty and full of run-off from construction, then the gator gar who'd been spawning in those bottoms for centuries might not venture up. After all, that's what happens when logging in the Northwest sullies streams; something overrides salmon homing mechanisms and tells them to give up. And the result of salmon giving up, along with other factors, added to a crash in 2008 and 2009 in which millions of fish disappeared from Canada to Mexico.

Still, I had to have proof that gator gar were spawning in Lollie Bottoms before I made a stand. So I contacted US Fish & Wildlife agent Lindsey Lewis, who said that he had netted a big fat female in there, and that he and Reid Adams had seen some adults above the concrete structure at the bottom of the bayou.

Then I contacted Reid. He said they were in there, but we really had no idea how important of a spawning ground it was, since we didn't have much data on that specific area. We both agreed that this would make a great study for some of his graduate students, and we both agreed that it was high time to start looking into this matter. He told me that if Conway Corporation was interested in working with the University of Central Arkansas on an environmental impact study, he would be more than glad to assist.

Thus, it was looking like another losing round of David vs. Goliath, which I was used to. When local cattle polluters up in Missouri illegally purchased the watershed land surrounding our source of drinking water for the town where I previously lived, I launched my crusade—since the water treatment facility there was incapable of filtering out all the fecal coliforms. There were plenty of municipal meetings, TV and newspaper interviews, independent research, participation in various forms of protest (some sanctioned, some not), and a riling up of the student masses. I even founded an organization focused explicitly on that issue, wrote an official letter full of state statutes and federal laws, and got a hundred other educators to sign it with me, then sent it off to the MO Department of Conservation, the EPA, and the Missouri Attorney General with an incriminating video. I generated so much explosive media in that corrupt one-horse town that it got to the point that I actually shut down City Hall. The arsonist mayor and her goon squad of drunken domestic abusers (the City Council) were so freaked out by a letter I published in the local paper that the mayor feared there'd be an uprising, so she canceled a public meeting with the people. Twice.

In the end, the state refused to step in. The cows continued to crap in our water supply, and four people and a cat were hospitalized for E.coli.

But now the terms were different. Now my city was attacking my fish, and wasn't even aware of it. So it was up to me to stick up for gar.

* * *

In the same sense that the once colossal Colorado pikeminnow (previously called "squawfish") are different from the smaller northern pikeminnows in Oregon, there are genes that get lost forever when a species is eliminated. When gene pools get wiped out, traits developed through thousands of generations also get wiped out. And those genes, of course, are good for something.

In the case of the Garhole population, the most noticeable characteristic is mangled tailfins. This might be caused by deformities, or maybe those gar just munch each others' tails up, or a combination of both. I currently keep two gar in a tank, so I can attest that they occasionally nip at each other, sometimes shredding fins a bit. Their tails always repair themselves, but in the case of the population in the Garhole, they don't always heal so perfectly. Whatever the case, I figure there's a sound biological reason for such deformities in the Garhole, which makes this population unique.

Photo 3. Mangled Tailfins of Garhole Gator Gar.

Photo 4. Mangled Tailfins of Garhole Gator Gar

Photos by Mark Spitzer.

But there's something else that makes these fish even more unique. For some reason, as far as we know, the alligator gar in the Garhole are all big ones. Nothing smaller than five feet long. To our knowledge, no new generations have supplemented this gene pool for at least twenty or thirty years, maybe even forty or more.

Obviously, I feel a connection with these fish. They're right next door, they're the gar I know best, and since it's hard not to get defensive when somebody messes with what you love, I was taking this matter personally.

Still, the argument that gar are bastard-fish wasn't going to get me much traction at the public meeting—even though gar are hardly considered *legitimate* in the eyes of most Americans. It's true that gar have been run out, shot up, dynamited, poisoned, thrown on the shores, and burned en masse to the point of extinction in many states, but it's also true that making such analogies wasn't going to help the situation.

I therefore resolved to subdue my urge to call upon the imagination of Conway Corporation and their appreciation for Shakespearian drama in order to make my case. Nevertheless, that's the approach that kept running through my mind as I sat there in my orange plastic chair, listening to the rationale for closing the old Stone Dam Sewage Treatment Plant. It was old, it was stinky, it was time to upgrade to a system that could handle the growth of the city. Also, the proposed site for the new sewage plant was downhill from all toilets in town, and gravity being a major factor in getting waste to move through tubes, it also made sense to have a sewage facility next to the river, where millions of tons of water could be recycled.

Another question that factored in was "What will happen to Lake Conway without six million gallons flowing into it on a daily basis through Stone Dam Creek?" And for me, living on Lake Conway, where I paddle out twice a day to check my lines in the cypress trees, I was invested in both places. Ultimately, though, I knew where I stood: I'd gladly see my lake take a hit, rather than the gator gar of Pool 7.

Unfortunately for the residents of Lollie Bottoms, who just had an airport forced on them for the seven corporate jets that sporadically fly in and out of the city, the proposed placement of the sewage plant was the result of years of strategic planning. Even more unfortunately, if Lollie Bottoms were to be zoned for industrial use, a slew of factories was expected to follow.

The attorney for Conway Corp. had prepared some type of environmental study. It wasn't an official impact study; it was a report designed to serve their purposes. The only species of concern in the area, according to the lawyer, were the lesser terns nesting downriver, which US Fish & Wildlife said would not be affected by construction.

I'd recently talked to Lindsey about those terns, and he said he had assured Conway Corp. that this species wasn't an issue. When I asked him why the gator gar weren't an issue, he told me that they never asked about fish. Plus, with the oil continuously spilling into the Gulf, US Fish

& Wildlife had been steeped deep in daily confusion. Almost all their agents were assisting in Louisiana, so any other concerns were on the back burner. Lindsey, in fact, had been down there all summer, and now it was fall. Millions of gallons had stained our Gulf shores, and at this time it was all about the clean up, which was drawing resources away from other problems.

To say the oil spill was "a distraction" is a monolithic understatement. This disaster had forced the Feds to take their eye off the ball, and Conway Corp. had benefited from this negligence. So when it came time for citizens to raise questions, I stood up for bastards with no hesitation.

This was no time for modesty, so I established my authority by announcing that I was a professor at the university, that I specialized in environmental topics, that I work on water quality issues, and that I wrote the book on gar. Then I dropped the bomb by explaining how there are alligator gar in them there bottoms that have been spawning there for centuries, and that state agencies across the South have been working with federal authorities to bring their numbers up. Then I added that UCA biologists were willing to assist in studying how this project can be implemented without harming the threatened fish or scaring them off, so that we can have the sewage system that we need and the gar won't have to be compromised.

I explained how alligator gar reproduction is a delicate process, and that quite a few private and public sector projects had been implemented to encourage gator gar spawning throughout our interconnected systems. I told them how our data shows that these spawns only take place on average once every seven years, how it's been documented that gar won't spawn if the conditions are not optimum, and how we still don't know enough about how turbidity and flow affect gar. Thus, what we were risking was kissing this particular gene pool goodbye.

I also argued that there's a concern regarding estrogen, which gets released into systems because of all the birth control that gets flushed into facilities that don't have the capability to filter out certain synthetic chemicals. Then, citing a 2006 University of Colorado study by Alan M. Vajda et al., regarding how male suckers in Boulder Creek were turned into females because of the release of synthetic hormones, I talked about how species all over the planet were going down due to too many mutant females in the mix—so we should consider this.

That's when the Conway Corp. lawyer busted in and said that the amounts of estrogen from the proposed sewage plant wouldn't be enough to affect the sexes of fish in the river. This lawyer, however, was no scientist, so I brought up the possibility of enough chemicals getting into the backwaters where gar and other fish spawn to have an effect. The point being: we needed a serious environmental study, not just assurances!

The silence that followed was definitely awkward. I truly believe that they had absolutely no idea that the continent's second largest freshwater fish—and an "imperiled" creature at that—was breeding in the place where they intended to process fecal matter.

Some other people spoke, and when the meeting was over, CEO Richard Arnold came over smiling smugly. He shook my hand in the spirit of sportsmanship and asked for Reid's contact info. I gave it to him and he said he'd follow up.

Then came the citizens and fishermen, who expressed their gratitude for what I had said and offered to do whatever they could. As I shook hands all around, I immediately knew what the difference was between this group and the executive who just told me he'd follow up on the impact study: these people, to put it frankly, were sincere.

* * *

The letter went viral within a week. It started in the city paper, got picked up by the state weekly, was reprinted in the state paper, and from there it migrated to all sorts of animal blogs, pet blogs, eco-blogs, and online newspapers throughout the state, region, country and world. *USA*

Today provided national exposure, and Animal Planet made this news international from their website in the UK. I was also interviewed on our local NPR affiliate, sticking up for gator gar.

Basically, the letter repeated much of the info I laid down at the meeting, but it also included a call for finding common ground on the spawning grounds by working together for mutual goals; i.e., the quasi-"beer summit" moment in which I stated, "It may be that this area is not vital to sustaining this population, and it may be possible for Conway Corporation to develop an environmentally friendly treatment plant." At the end of the letter, I added the information that "Conway Corporation is currently collecting comments from the community for the next ten days regarding this project. I'd like to encourage everyone with an interest in conserving this important natural resource to write to CEO Richard Arnold . . . as soon as possible and request that this matter be seriously studied before any construction begins."

After that, the most common question I heard on this matter was "Can't we get them listed as an endangered species?"

The answer to this was complicated. I'd received an email from a Little Rock attorney a few months back who had read *Season of the Gar*. Like most of those who contact me after reading that book, he wanted to know what he could do for the cause. But unlike most of that bunch, he was all fired up and angry as hell.

He told me his story. He used to be a young and passionate environmental lawyer who worked on endangered species, so he knew how to get the job done. Now, however, he had metamorphosed into a fat, lazy, corporate type who didn't do anything good for the world. So his life lacked meaning, which bummed him out, because he still loved fishing, and he felt he owed it to his son to help preserve our natural world.

Hence, he was ready to go *pro bono*, start filing paperwork, and get gator gar protected across the country. He'd work on the weekends, he'd work at night, but whatever he did, he'd do it right—so he wanted to know what I thought about that.

I was at the International Gar Conference at that time, so this topic was convenient for starting conversations with the experts. I spoke to all sorts of authorities, and here's what I found out:

For the first time in history, science now has a coordinated effort going for gar. As mentioned before, state and federal agencies are working with private and public entities, and the hatcheries and fish farmers and commercial fishermen are in on it too. This last group, though, is the biggest reason behind the logic for not making alligator gar off limits across the board. As Alysse Ferrara told me, commercial fishermen are providing specimens and collecting data, which is one of the biggest boons to this niche of biology that's ever occurred. And as Lindsey Lewis explained, if gator gar get listed as endangered, then labs won't have access to them. And presently, we're learning a lot about aging, growth, egg generation, DNA, the whole shebang—so it would be detrimental to our momentum to suddenly lose all that. Plus, there were just too many big ones holding out in Texas and Louisiana. Sure, there may only be a few thousand alligator gar left in the country, but for a federal act to protect these fish, there'd have to be a whole lot less.

So that was that. The Little Rock lawyer was disappointed that I couldn't give him the go-ahead he was hoping for, but he respected the evaluation I offered him.

The residents of Lollie Bottoms, on the other hand, were up in arms and getting desperate. In addition to the airport imposed on them, they also had an uninvited soccer complex burning bright with light pollution, and it looked like there would soon be convoys of heavy equipment barreling by as their property values dropped like lead zeppelins. Also, there were health concerns, issues of imminent domain, and of course, no one wants to smell that shit. Literally.

Their leader came to me a few times. She called me at the office, she called me at home, and every time I listened. For half an hour or forty-five minutes I'd listen. I could never get off the phone with her—to the point that "people" became an annoyance, even though the individuals involved were definitely for something I was for.

This caused me a great deal of consternation. On one hand, people were coming to me and asking for help—so what kind of a guy would I be if I told them no? On the other hand, my area of study was fish—not civil rights. Which is why I ultimately decided not to stick *up* for gar, but to stick *with* gar.

Meanwhile, the ten-day comment-collection period went by in a flash, and during this time, the chat-room conversations following an online garticle in the local paper shifted from practical to not very helpful. As usual, whenever the *Log Cabin Democrat* focused on gar, people started talking smack. Some complained about how ugly gar are, but then there were the know-it-alls claiming there ain't no shortages. One old timer chimed in, claiming that alligator gar were all up and down Cadron Creek (when those are longnose) and should be shot on sight and left for dead. When I replied with a quip about how this type of ignorance is what stereotyped this fish in the first place, he snapped back that he'd been fishing these rivers for fifty years and wasn't about to listen to no professor who sits at a desk and doesn't know jack about being in the field.

I've come to expect these responses. There's a prejudice against gar that's been part of our culture for centuries. It's the same type of bias that used to just exterminate folks, but now that this is unacceptable, it's safer to just round 'em up and ship 'em back to where their different looks and different ways won't mix with ours. I encounter this all the time with gar: throwbacks so stubborn that they refuse to consider evidence that can put their opinions to the test. But luckily, this generation is dying out, and the generations following it are a lot more tolerant. At least we've got that going on—while the world goes to hell in a hand basket.

But back to the comment-collection period, which I hoped would be open and honest. It wasn't. Richard Arnold claimed he didn't receive any letters, even though I had sent one in myself and I knew others who'd done the same. The result being: Conway Corp. moved one step closer to building their "state-of-the-art" sewage plant.

Still, there was one encouraging factor: now that the project was moving ahead, a City Council Zoning Commission meeting had been scheduled to discuss the granting of the permit for construction. Technically, this would delay the project. And so the war slogged on.

* * *

The press release I sent out said I'd be challenging Conway Corporation at the upcoming municipal meeting, and the media glommed on, looking for a fight. The Gar-nut vs. Goliath theme was pumped up, the *Arkansas Democrat-Gazette* sent a photographer up from Little Rock, and they ran an article billing the show like this:

> The city of Conway will hold a public hearing at 7 p.m. Tuesday at the District Court building to discuss planning and zoning for Conway Corp's proposed sewage-treatment facility . . . Mark Spitzer, a professor of writing at the University of Central Arkansas in Conway, plans to be there, and Conway Corp CEO Richie Arnold isn't surprised . . . Spitzer really only has one concern: gar.

Up in Fayetteville, the *Razorback Reporter* also ran a story, noting "UCA biologists have volunteered to help Conway Corp to run tests on the area to ensure the gar population is unharmed . . . [but] Arnold has not heard from any biologists at UCA." So after asking me for contact info, Conway Corp was spinning it as a situation in which the university refused to reach out to him.

Then the showdown:

When I walked in, I was greeted by the residents of Lollie Bottoms. They were disabled, in wheelchairs, had problems breathing, and they were looking for me to present their case—because I had clout (or so they thought). I, however, was only prepared to talk about gar.

It didn't matter, though, because they did a pretty convincing job. One by one, they addressed the Zoning Committee, expressing their concerns regarding their safety, their quality of life, and the fact that most of their homes had been there for generations. Even kids addressed the committee. They'd worked hard with their parents to actually physically build their homes, and now a big old stinkbomb was moving in. Where's the justice? Why punish us?

Richard Arnold responded with the NIMBY (not in my back yard) argument. He said that wherever these facilities get constructed, there's always resistance, so it's unavoidable. Then he swore that the plant wouldn't stink.

He had come up to me before the meeting and shook my hand with a big grin. How that guy shaves in the morning, I don't know, but I know I was seeing this conflict from my own biased point of view. To me his agenda was nothing more than creating waste in the name of waste. Ironic or not, that's what I saw going down. This guy, so it seemed, was as greedy and self-serving as King Lear's slutty daughters.

Well, okay, maybe they weren't slutty; but they were definitely those other things. So no wonder Lear raged to the fool upon the hearth. That's what we do when we run out of hope, and that's what I was starting to do.

But I still had some fight left in me. So when I got up in front of the Zoning Committee, I decided to stick up for the people. I talked about the legality of this issue, how both public meetings were announced in underhanded ways designed to keep citizens out of the process, and how Arnold's claim that no one had responded to the call for public comments was total B.S. In essence, I was calling him a liar in front of everyone. But "Richie," he just shook his head like I was some sort of whacko.

This tactic wasn't working for me. My voice was starting to stammer, and it was clear that my comments weren't as articulate as they could be.

So I switched it to gar and talked about how two years of construction would muddy up the bayou, mess with the flow, and keep gator gar from spawning there.

All in all, it was a disappointing sparring round. I threw a few lame punches, and then stepped down. The featured event had been a flop.

But then a young state senator named Rapert stepped up, representing the constituency of Lollie Bottoms. And when he did, he brought a lot of "street cred" with him. He spoke slowly and deliberately and got his message through in a manner that was way more effective than my lousy attempt to slap Conway Corp. on the wrist. And I was impressed. Impressed that I could be impressed—by a Tea Party Republican.

Like most Americans in this severely stressed-out bipartisan climate, I tend to see those who don't agree with me as *the opposition*—which can sometimes be a mistake. It's important to remember that even if Democrats and Republicans and Independents don't agree on particular subjects, common ground can still be found.

That's the way it was in Missouri. It wasn't just Democrats fighting for clean water, it wasn't just Republicans. I worked with members of both parties and we lost the fight together. Still, it wasn't a complete loss, because six months later another issue arose. The pornographer who owned the hotel downtown had installed a giant spotlight on his roof which shot obnoxious beams of rotating light into the night sky. He saw this as advertising, but when I was six miles out of town and fishing at night, those damn lights were washing out the stars for over 25,000 people in the county.

Another round of concerned citizens raising complaints ensued. Since we were already united by the issue of cryptosporidium contaminating our water supply, it only took a few months to force the new City Council to amend a law about what types of lights can be used at night.

In that case, the Us-vs.-Them conflict worked out in favor of the people. On the most part, though, Us vs. Them is hardly ever constructive. Still, that's the situation I was now in. Even more unfortunately, that's the situation the residents of Lollie Bottom were in as well as the gar. What we needed was something to bring us together, something to put us on the same side, so we could approach this matter from a mutually beneficial angle. That, of course, is the ideal, but with seventy million dollars at stake and America's fugliest fish in the middle, I had no idea how to go about that.

Anyway, the Conway City Zoning Committee—much to the chagrin of Conway Corporation—decided to postpone their decision until all members of their team could visit a similar sewage plant in Little Rock to see if it stunk or not.

When I walked out, the last thing I saw was Richie Rich smoldering.

* * *

The way it all went down is that the Zoning Committee took a trip to the northwest corner of the state and saw a different type of sewage plant that didn't stink, so made the decision to zone the proposed section of Lollie Bottoms for industrial usage. This news coincided with my three gator gar fingerlings going belly up in my tank. I'd been feeding them tiny minnows and larvae for the last few months, and they'd grown to about two and a half inches long. Summer, though (which includes fall in Arkansas), is a difficult time to keep fish. With the warmer water, bacteria feeds voraciously, and infections are common, especially when you're moving fish from tank to tank with dip nets. All it takes is one adverse microorganism and the next thing you know somebody's swimming at an obtuse angle. After that, it's fungus or body slime or eye slime or one of a hundred other fuzzes which can decimate an aquarium.

But it was a fitting end to my clash with the Titan. It's what I expected, and in a way, it was appropriate to view the death of these three Arkansas alligator gars as an omen of things to come.

Still, as is my nature, I resisted the urge to agree with the Shakespearian character of Edgar—who, at the end of *King Lear,* proclaimed, "The weight of this sad time we must obey."

Meaning rather than suck it up and lump it, I decided to do what we all do when we can't accept the situation. I chose denial. Denial that my city would create such waste to process waste. Denial that those rare and ready spawning grounds won't be there in two years' time. Denial that I couldn't shut down City Hall or even keep three gar alive.

But as Lear howled to the cataracts and hurricanos, "Blow, winds, and crack your cheeks. Rage, blow . . . spout / Till you have drenched our steeples, drowned the cocks . . . spill at once" (III. ii.). Because the more rain we get, the more deluvianed land we'll have for gar to take back what they once had eons before us fools came along, or even had a pot to poop in.

After all, gar were here first—before prehistoric man, before the Native Americans, before the fact that if humans want to treat their waste, something's got to give. And in this case, gar lost.

But that's what they're used to, and that's why they've been around so long. As Edgar also states at the end of Lear's tragedy: "The oldest hath borne most"—a disposition which makes gar, who are masters at adapting, even more fascinating, and difficult to just flush away.

Chapter 3

Finding Judas

The True Meaning of "Fishing Support"

After an entire year of getting skunked (except for a few cats), the Fishing Support Group was more than discouraged. Still, "FSG," as we also called ourselves, was not that discouraged at being discouraged, since we'd designed our joke of a group as an excuse to drink beer on the river while our wives participated in some sort of writing club. Whereas the wives denied that they were workshopping each other for therapy, us guys embraced the idea that we were going out once a week for the express purpose of providing each other support for angling angst that the ladies couldn't empathize with.

So that's who we were: Minnow Bucket, T-Bone, Ty-Stick, and Hollywood (the latter being me, since I'd been on TV defending gar). Those were our fishing nicknames. Suffice it to say, we were three professors and a payroll specialist—but not on Thursday nights.

Fishing Support Group was mostly active in the summer. We'd go out when the day cooled down, bring minnows, worms, sometimes livers, our fishing poles, and a bunch of jug lines (mostly plastic vodka bottles strung with circle hooks and railroad ties as weights). We'd take my 1959 Whitehouse Runabout (a.k.a. the *Lümpabout*) that I salvaged from a lake in Missouri, and we'd launch on the Arkansas, usually below the Toad Suck Dam.

After throwing out our jugs in various coves, we'd anchor somewhere and toss out some bait. Our goal, of course, was to catch an alligator gar—but that being highly unlikely, the expectation was to hook a few catfish at least.

When I'd get home, my wife would ask what we talked about, and my answer was always "fish." Because that's why we were out there, and that's what we mainly discussed. Not our emotions, not our relationships, not the "support" we pretended to need.

In the summer of 2011, FSG talked a lot about gar, the main reason being that I was obsessed with gar and had been going to the Garhole every month for over a year to observe what was going on. And what was going on was drought, which kept the tributaries low. And because of the lack of flow, I'd hardly seen any gator gar.

The experts I knew figured they were out in the main river. In fact, I'd seen a few big ones rolling behind some jetties, and Minnow Bucket had recently seen a few down in Little Rock, just swimming along under a bridge. The ones he'd seen were between three and four feet long, the same size of those I had spotted—which was a good thing. It meant that a new generation was coming of age, and that their numbers were probably higher than usual, since you typically only see the tip of the iceberg.

Fishman, also, confirmed my theory. He was the only commercial fisherman left in the Toad Suck environs, a grizzled veteran of the area who'd been running nets for flatheads and buffalo for more than thirty years.

Whenever I'd stop by to get some gar bait from him, he'd load me up with flayed spines, then insist I don't pay a dime. He had read about my activism on behalf of gar in the local newspaper, and he was all for me doing what I was doing if it could help the populations thrive. "Just keep working on them gar," he'd tell me, then throw in some paddlefish steaks or an American eel as an incentive.

As Fishman explained to me, he wanted to see the gator gar populations in Pool 7 and Pool 8 return to what they used to be. Since he'd fished this river all his life, he knew the big ones used to be bigger, and he knew it was due to the apex predator in the system, which used to help balance it better when there were more.

In the last few decades, Fishman had caught more alligator gar in the Arkansas River than anyone else on the planet, the largest being eleven and a half feet long—so he claimed. He'd caught it back in the nineties and it was so damn mongo that he couldn't haul the gillnet into his flatbottom boat. So he towed that sucker over to a sandbar, where he and his wife tugged it up on shore. They got it in the boat, took it to the junkyard, and it weighed in at 328 pounds. But since he'd caught it on a commercial license, he couldn't claim the state record—even though the new official world record is 327 pounds and was caught on a commercial license.

"There's one out there now that's just as huge," Fishman always told me whenever I came in, "and it's an albino!" He told me where it lived, and he told me he was going to get it, then build a thousand-gallon tank right in front of his fish market so people could come and check it out.

Mostly, though, he'd tell me about the smaller gator gars. The three- and four-footers he was catching and sometimes keeping. At first this freaked me out.

"But there's only a couple hundred left," I told him, referring to Pool 7. I was thinking of the old school population that lived in the tributary Team Gar and I knew so well. That population is estimated to be between seventy-five and a hundred, and I was factoring in another possible hundred that stuck primarily to the main river. With fewer than a hundred actually tagged and accounted for in the state, UCA biologists (according to Uta Meyer's August 30, 2010 article on the Mother Nature Network website) estimated there to be about five hundred in Arkansas.

Fishman, however, just laughed off my estimate. "There's hundreds out there," he replied, referring to the stretch beneath the Toad Suck Dam. He then showed me his records, since he was required to report any gator gar he caught to Game and Fish. And he did.

It was the beginning of summer at that time, and looking at his graph paper, where he wrote down sizes and weights and where each particular gator gar was caught, it looked like he was catching about one per week: forty-six inches, forty-eight inches, forty-two inches—but nothing over five feet long.

"When I catch 'em bigger," he told me, "I let 'em go."

"Have you ever caught a tagged one?" I asked.

He replied no, which basically meant that the gar he was catching weren't part of the tributary population. Since half of those from the Garhole (forty-seven to be exact) had already been tagged over the last five years, it was curious that Fishman hadn't caught any. This was also a good thing.

Fishman would tell me where to go, and I could see from his records where he'd caught them. But when I went to those spots with FSG, we never got squat.

So at the end of August, we got together, the entire Fishing Support Group. It was the last time we expected to fish together that summer, because school was starting next week and there'd be classes to plan on Thursday nights.

We launched in Pool 8, which we hadn't fished much, so I wanted to give it another shot. Lindsey Lewis had been scouring this less-developed stretch for years, and he figured that the alligator gar population in it was comparable to the one below it. In fact, he'd recently told me a story about some trotline fishermen he'd met up by the paper plant near Morrilton a few years back. There's a creek up there, where the trotliners found nine giant gator gar hooked on one of their lines. According to Lindsey, they let them go.

Anyway, we found a jetty and threw out a bunch of jugs, then motored to the cherry spot—which I'm not about to reveal. We didn't know it was the cherry spot at the time, but we were about to find out why.

Like usual, Minnow Bucket rigged up with a sunfish, which he'd caught down in Little Rock on his lunch break. Minnow Bucket worked at City Hall, which was right on the river, so he'd go out with his cast net, catch some small minnows or shad, then catch panfish or bass with those. That's just the way he rolled: using small fish to catch bigger fish, then bigger fish to catch even bigger fish.

Meanwhile, I threw out a worm and a liver, Ty-Stick threw out a minnow, and T-Bone threw out a goldfish—which are commonly sold as bait and legal to use in Arkansas. This is problematic, since they have the potential to breed as invasive species. Other states have laws against using goldfish as bait, but they sure stay on the hook a lot better than shiners do.

We cracked our beers and the dusk darkened. It was nice to be out on the river not talking about our feelings, and it was cool to cool down after another hundred-degree day as we resigned ourselves to not catching anything.

Around ten o'clock, I suggested we try a new spot. We were feeling lazy, though, so decided to stick it out ten more minutes before hauling up, firing up, and cruising to another spot.

That's when Minnow Bucket said he had a fish, and by the bend of his pole, we could see it was a big one. He was hoisting it up and it was resisting. Probably a channel cat.

Then something leapt forty yards downstream. It was totally dark out there, but we could see an eruption on the surface and a silvery blur arcing above it. It then splashed down from where it came, creating another explosion on the water.

None of us knew what it was and Minnow Bucket kept pulling it in. When it got about twenty yards from the boat, it leapt again, straight out of the water. Four feet out of the water, in fact. And this time we all saw it. It was a long-ass, thrashing tube, perhaps a longnose, once again splashing back in.

So of course we whooped as Minnow Bucket kept on cranking. He was bringing it in, our adrenaline was surging, and then it started heading toward us. Then passing us, straight and fast, cooking along like gar tend to do. Not that we could see it, but we could see where the line was heading: across three of our other lines—like a torpedo. Which was another strong indication that we had definitely hooked a gar. So I grabbed the landing net.

Soon it was beneath us, and Minnow Bucket was bringing it up. And bringing it up. And bringing it up.

The shape formed, broad and elongated. Its silhouette lightened, turning grayer. Our headlamps were shining down on it, and then we saw that wide blunt beak giving it away. It was an actual alligator gar!

Holy Crap! The gar took off and made another run.

FSG members claim that when I saw that fish, I threw the landing net over my shoulder. I can't remember doing this, but I do remember thinking, *There's no way this dinky thing is going to get a piece of that!* According to my friends, the net came down in the boat.

That's when I started shouting orders. I told Ty-Stick to open up my tackle box and get out the yellow stringer. He did it, while Minnow Bucket fought the gar back to the boat and brought it up parallel.

This one wasn't going to get away! Ty-Stick handed me the stringer, I reached down in the water, and the gar was way too beat to object. I wrapped that stringer underneath the pectoral fins, then ran the metal pokey part through the eye on the other end, tightened up, and noosed that fish under its armpits.

Having seen Captain Kirk do this on the Trinity, the method proved effective and as natural as anything I'd ever done. I coiled the un-fish end of the stringer around my wrist three or four times, then yanked that gar right into the boat.

For a few seconds none of us could believe it. We actually had it! It was ours! This was definitely a bonus!

Out came the cameras and flashes went off. We were cutting free the tangled lines, yahooing our heads off, and looking for the measuring tape. But we had no measuring tape. We could tell, though, that this gar was just as long as my four-and-a-half-foot paddle.

In the meantime, Minnow Bucket was taking long measured breaths, as if he was hyperventilating. He wasn't—but he was breathing hard, and he was deliberately trying to keep his exhalations and inhalations even. Because when you catch an alligator gar, your heart starts jackhammering. I've felt this before, and he was definitely feeling it now. He had met the Objective! He had actually caught the elusive Big One—which guys like us fish for all their lives.

Photo 5. Hollywood and Minnow Bucket with Alligator Gar.

Photo by Tim (T-Bone) Thornes.

This missile of a fish was smiley and steely and sharp. It was a beautiful juvenile gator gar, just a few years old, and it only weighed twenty-seven pounds. We were a bit disappointed that this fish didn't top thirty pounds, but we weren't about to look a gift-gar in the mouth. Still, we did look it in the mouth and saw two rows of razor-sharp fangs running along the upper jaw-line.

We let it go as fast as we could and exchanged high fives all around. This was a cause for celebration, so we cracked another round of beers. But nobody fished. That part of FSG was over. Having had more than our fair share of fun, we were more than glad to just exist.

But the evening wasn't over. We still had our jugs to pick up, so we motored over to the jetty and started pulling them up. We had one small cat which we tossed back, and then we found a displaced jug with nothing on it. Something had severed the line and the tackle was gone. Minnow Bucket's thirty-pound braided line was frayed as if something had rubbed right through it. And there's only one type of fish that we were aware of that was capable of pulling a stunt like that.

Then we found another jug with a line that had also been lacerated. Even snapping turtles can't bite through this type of line—so something was up in this neck of the woods.

Especially us!

For the next three days, we couldn't stop buzzing. It was all about the photos we were sending each other and the continuous rush that wouldn't fade away. It was all about that sexy, fantastic, miracle fish that had graced us with its Gar Glory! Because when you catch an alligator gar, you come face to face with God.

* * *

It was the only known alligator gar caught on rod and reel in the state since the new gar laws had been passed in 2010. It might've also been the only known gator gar caught on rod and reel in Arkansas since Alvin Bonds of Clarksville caught a 215-pounder on this river in 1964. That, of course, was following the crash at the end of the fifties—back when alligator gar used to rule the White, the Red, the St. Francis, and just about every non-mountain Arkansas river from Oklahoma to Tennessee.

Minnow Bucket had reported his catch to the Game and Fish Commission the very next day, as is the requirement for the privilege of downloading a free alligator gar permit. When the new rules for gator gar were designed, Game and Fish decided to rely heavily on information from bowhunters and fishermen. In fact, tournament shortnose shooters had been invited to have a voice in drafting the regulations, since they're the ones who see gator gar the most. Game and Fish's idea was to make use of humans in order to find and track various populations of gar, an idea that was now starting to pay off.

Our "support group," however, never really realized that when we signed up for those gator gar permits we were pledging to provide fishing support in the form of data. Since we were obliged to pass on information, the real fishing support we were providing was logistics on where gator gar were. As planned, this information would be used to pinpoint possible hot spots so that the US Fish & Wildlife Service and the Arkansas Game and Fish Commission could be of service and serve us. *Us* being not just humans, but alligator gar as well.

But when Minnow Bucket called Game and Fish, they were totally unprepared to process this information. Nobody at the 1-800-report-a-gar-hotline knew what to do, but eventually the news got through and spread to Lindsey Lewis. Within an hour, he called Minnow Bucket up, and Minnow Bucket gave Lindsey the lowdown.

It was the info he'd been looking for. Lindsey figured that if he could get just one fish from that population and attach a radio transmitter to it, then that gar would lead him directly to the wintering hole in Pool 8, which he'd been trying to locate for over a decade.

So Lindsey got to work. And two months later, he cced. me this email:

> Last Friday we went out just upstream of . . . the area where Mark Spitzer and his friend caught an AG over the summer . . . with the intention of catching and radio tagging at least one fish. We were successful. This fish, which we named Judas, will hopefully, finally, lead us to the wintering hole(s) and other fish over the winter in Pool 8. Below is the data on the fish along with some pictures.
>
> As always this was an incredible cooperative effort. We used reconnaissance information obtained by Mark to narrow down a good location; Chris, Tommy, and I put out the nets and performed the tagging; and the tags were obtained and loaned to us by Lee Holt. Thanks all and good work!

Species/Id: Alligator Gar (Judas)

Location: Pool 8 (0.5 mile upstream of [REDACTED])

Date/Time: 10/21/1113:30

Crew: Lindsey Lewis, Tommy Inebnit, Chris Naus

Lat/Long: [REDACTED]

Radio Tag: 150.250

Floy Tag: FWS109

PIT Tag: 4A4566004F

Length: 137.5cm (4.5')

Weight: 18.13 kg (40 lbs.)

Genetic Clip: Yes

Weather: Sunny, Clear, Calm

Air Temp: 76F

Surface Water Temp: 66.4F

1m Water Temp: 66.4F

Flow: None

Depth: Variable 12-50 ft.

Structure: None

Water: Clear

Observations: Lots of surfacing longnose gar, sonar location of
numerous (dozens) 10-50 lbs. gar in area

Gear: 600' gill net tied together (2) 4" mesh and (1) 5" mesh

Notes: Fish was caught in the 4" mesh near the top of the net,
gilled, but mostly entangled. Net was anchored on one end in
~20' of water and used like a seine slowly pulled by the boat in
a swinging door maneuver over the area with surface and sonar
observations in 30-40' of water. There was one hump immediately
adjacent to the surfacing fish that was 12" in depth and around
20" in diameter. The jug closest to the fish immediately began
bobbing deep into the water when the fish hit the net and was
caught. We stopped seining, anchored the net and retrieved the
fish. No other fish were caught and no additional effort was made.

The fish appeared to be in perfect health in the boat and . . . it
responded well when released by quickly diving with force. The
radio tag signal was detected and confirmed to be functioning
following release.

I looked at the photos of Judas, who'd been commissioned, in a sense,
to betray his flock. I'm sure that's why Lindsey named it that—even
though the original Judas hadn't been such a bad guy. The "Gospel of
Judas" had been discovered in Egypt in the seventies, and had been
translated and published in 2006. This news is still getting out, and hasn't
yet registered on the consciousness of the masses, but as it turns out,
Judas was only obeying orders from Christ—to make it look like he'd
ratted him out. Archeological experts confirm this, and historians deem
this document to be authentic, according to *National Geographic*'s May
2006 article entitled "The Judas Gospel." Point being: the biblical Judas
was no Benedict Arnold, and Lindsey's Judas could very well be a new
kind of Moses leading us to the Promised Land.

Analogies aside, though, Lindsey's Judas had been caught in the same
place as Minnow Bucket's gar. Not only that, it was the same length, but

thirteen pounds heavier. Since gar are territorial, and since they typically fatten up for the winter, and since Minnow Bucket's gar had two months to chow down, this meant that Minnow Bucket's gar and Judas could very well be the very same fish.

Lindsey's gar, however, was spottier than Minnow Bucket's fish, but as I remarked in my response email, gar are often paler at night than in the day. I'd learned this by watching my own pet gars, who become a lot more ghostly-looking when the aquarium lights are off at night.

But chameleon or not, we'd also had our jug lines shredded in that spot. Plus, a week before Judas was caught, Minnow Bucket had been trying his luck from shore in that spot when he got snagged on something. Then that something started slowly moving off, before snapping his thirty-pound test like it was thread. Lindsey had reported big longnose in that area as well (which tend to hang with alligator gar), and had seen some sort of big "hump" that could've been one of Judas' buddies.

Most likely, Judas wasn't Minnow Bucket's specific gar; it was probably one from the spawn of 2007, which had been the last successful spawn we knew of (documented in Tommy Inebnit's 2009 UCA thesis). Since Fishman had been catching them this size in this section of the river, and since Minnow Bucket had seen some of similar dimensions down in Little Rock (as had I in Pool 7), it was looking like what we were seeing now was a new generation of alligator gar coming on strong in the Arkansas.

Anyhow, I sure hoped this was the case. If it was, that meant I was one lucky bastard: to live right next door to the most thrivacious population of gator gars in the state when that's what I'm most interested in investigating, researching, and catching on a big old bloody drum. Meaning things were looking up for me just as much as they were for gar. But most of all, things were looking up for the state.

Because basically, we're now getting a second crack at being good stewards of alligator gar. Whereas we failed before, the science and history are now in perspective. And because of this, there's a greater awareness of what this niche means to its system as a whole. So as the fracking world goes to crap with oil spills and climate change, not to mention the usual toxins polluting all our waterways (mercury, lead, PCBs, pesticides, etc.), our gator gar are also getting a second chance to regain their natural habitat. But what's remarkable about all this is that even if us humans drown in the floods we're too chicken to even envision, the Arkansas alligator gar will continue on as members of one of the oldest living fish families on the planet—resurrected with some help from Judas.

Photo 6. Chris Nau and Lindsey Lewis with Judas.

Photo by Tommy Inebnit, Courtesy of Lindsey Lewis.

Photo 7. World Record Gator Gar.

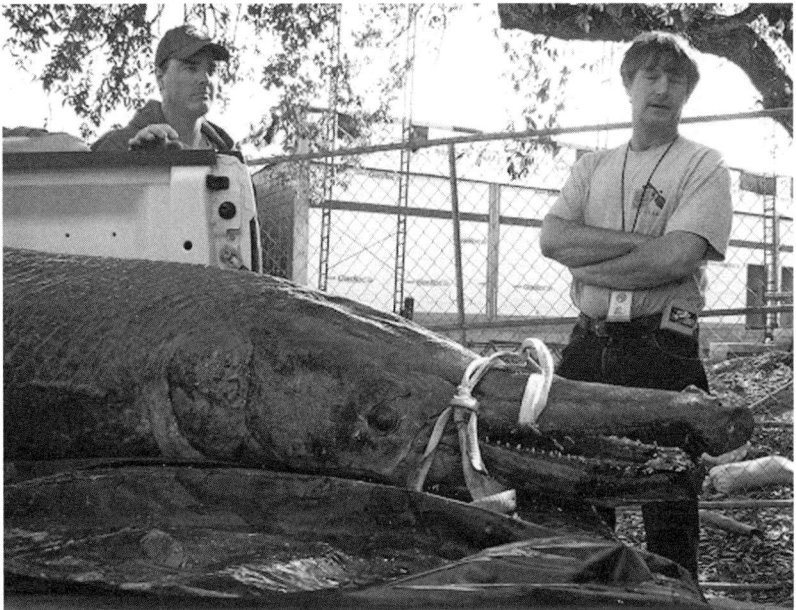

Photo Credit: Ricky Flynt, Mississippi Department of Wildlife, Fisheries & Parks.

Chapter 4

Enter the Next Generation

For the first time in my life, I felt confident enough to claim that things were looking up for gar, due to the fact that they'd made the mainstream. In Mississippi, a commercial fisherman on Lake Chotard had hauled in a world record alligator gar tangled in his net. It was eight-feet five-inches long, forty-eight inches in girth, weighed 327 pounds, and the images had instantly gone viral. Also, a federal jury in Texas had just convicted a fisherman in an international gator gar sting, and the news had spread like wildfire. The guy caught four four-footers on the Trinity River and transported them to Florida where he kept them in a swimming pool. This gar-monger then sent them off to a corporation in Japan and ended up busted for illegally transporting fish across state lines. Plus, millions of viewers of *Hillbilly Handfishin'* had recently witnessed a gar in Oklahoma attack a city slicker's shiny pendant, flashing like a minnow. The show then cut to some stock footage of an aquarium gar, to let us get a gander at one. The fish they showed, however, was a saltwater Atlantic gar (a.k.a. needlefish), which is totally different from freshwater gar in the US.

What all this attention said to me was that our communal interest in the granddaddy of the species was increasing in direct proportion to our efforts to build their numbers up. Of course, there was no way I could verify this, but I knew there was a correlation between sustaining alligator gar populations in order to provide sacrificial goats to spotlight in the media and the fact that we actually had some big ones to spare.

In a way, gator gar had become a phenomenon a lot like noodling. Here's what I mean by that: Ten to twenty years ago, it was common to hear stories about guys mucking into streams, where they went "hogging" for big old catfish with nothing but their bare hands. For the most part, these tales were just unconfirmed rumors in the consciousness of pop culture. But after the documentary *Okie Noodling* came out in 2001, there was an explosion of cable shows that shattered the idea of noodling being a myth and established it as something real. In a parallel sense, these shows brought gar out of the closet and made them much more tangible to the American couch potato.

But for Team Gar, behind the scenes, we still had a lot of work to do. Fortunately, it was work we loved so much that we'd be willing to pay to do it. We didn't have to, though, because it was all about getting up at 5:00 a.m., then staying out until after sunset, gillnetting gar.

Like usual, we met at the USFW parking lot: our leader Lindsey Lewis; ichthyologist Reid Adams; reformed gator gar-slayer Jaypea Atkinson; gar-students Chris Nau, Loren Stearnman, Casey Cox; and me. It was a cold, dark January morning, and we were heading out to the Garhole, which had become an annual pilgrimage. Our mission: to net 'em, tag 'em, collect data, and let 'em go.

Still, I didn't have much faith that we were going to get any. Two weeks before, I'd gone out there with my wife when the water was an unusually high fifteen feet at the measuring point upstream. For the last two years, the average reading had been five feet, even though the depth in the Garhole can be as much as fifty. I'd suspected that because of the recent rains, the gar would sense a change in flow and return to their traditional wintering hole. And I was right.

They were rolling all around us, sometimes five or six at a time, slapping the surface in all directions. I recognized the longnose no problem, but as for gator gar, I could never get a good glimpse. Still, there were a lot of four- and five-footers porpoising all over the place, sometimes thumping against the hull, those spatulated, spotted tails waving goodbye as they shot back down.

It was an extremely warm day, and this flurry of activity was definitely irregular for the winter. Usually, they just hunker on the bottom, semi-dormant and hardly ever rising for air. My own pet gars do this as well, their metabolism slowing for the season. But those Garhole gars, they were burning calories like crazy.

Robin was shooting video of the splashing action, which picked up every time a cloud eclipsed the sun. I've been observing this type of behavior for years, and have yet to see any other species get so riled up when a front pushes through, thus, my conviction that gar are highly sensitive to barometric pressure.

Anyway, my two years of coming out here and not seeing much were now paying off. When you're in the middle of a thrashing garfield, it's a visceral rush, one I was glad to share with Robin. She had come out here with me half the time, so the spectacle surrounding us was finally paying off for her too.

I came out the next week with Minnow Bucket and the roiling I'd witnessed with Robin was gone. At that point, the depth was down to seven feet, and the only fish we saw were the two blue cats we picked up on our jugs motoring back to the launch.

But now, driving through the dawn with Team Gar (two big trucks towing two big boats), the water was back to five feet deep. Lindsey was certain that they were still in there, and he was certain that they were hanging out together as gar tend to do in the winter, despite the fact that in the past two years the big ones had spread themselves throughout the system.

I was resigned to getting skunked, but I was also along for the ride. Especially since I had this idea for another gar book: a sequel to *Season of the Gar*, to be entitled *Return of the Gar*. That's right, I had the name already picked out, even though I still wasn't sure what I could say that I hadn't already said. What I was sure of was that Team Gar always talks gar, so I knew I'd get some new information on the ugly fish we all love.

By ten o'clock we were on the river, and by 10:30 we were at the far end of the Garhole, laying the downstream block-net in the bigger boat. The "running boat" was upstream laying the other block-net, so any gar in between would basically be trapped.

We'd done this a year ago, but only managed to catch one six-foot longnose that weighed thirty-five pounds. It was a hell of a fish and could've been the state record if caught on rod and reel. Lindsey took out an ultrasound machine, which was capable of detecting eggs in gar. If he found eggs, that meant the fish was a female. And in that particular longnose, Lindsey detected a bunch of eggs. So now we had a harmless way of sexing fish in the field, which was a revolutionary improvement over the old method of cutting them open to see if they had gonads. Not only did this mean that we could gather basic data more humanely, but it also meant that the popular hearsay about all large gator gar being female was being debunked.

The year before we caught that longnose, we'd caught fifteen alligator gars in this spot ranging from five to seven feet long. But this year, nothing was rising, not even any pipsqueaks. Like I told Lindsey earlier, we missed our window.

Nevertheless, we spread some nets between the block-nets, then went tearing around to stir them up. Nothing happened, so we broke out our sandwiches. Everyone except Reid, that is, who brought his traditional Vienna sausages in a can—which, being from Alabama, he pronounced "vye-enna."

Then we saw a float bounce. Something was caught in one of the nets, and then we saw a shark fin rise. But it wasn't a shark fin; it was the tail of a paddlefish, which we soon untangled. It was five feet long from head to tail and upwards of thirty pounds—a cartoony looking creature with a big old shovelnose. We took pictures of it, but as I found out later, my old fashioned .35 mm. had bit the dust.

Photo 8. This Isn't the Paddlefish We Caught, but It's About the Same Size and Definitely Worth Showing. This One Got Chomped by a Bull Shark in Louisiana.

Photo Courtesy of Lindsey Lewis.

Releasing the fish and finishing lunch, it looked like we'd be shutting down by noon. If there were gar in this hole, they would've rose by now, because that's what they do every forty minutes or so.

Lindsey, meanwhile, took off upstream in the running boat. He had a fish finder on that craft and was adamant that they were in there somewhere.

Twenty minutes later, the running boat came screaming back. "They're right around the corner," Lindsey told us, "hundreds of gar!"

As it turns out, Lindsey's fish finder was capable of taking pictures under water, and he sent me some JPEGs later. In one image there were ten silhouettes which were clearly longnose, but then there was this ghostly figure slightly defined by a frosty shadow. It was an alligator gar so wide and blunt you could see the shape of its primeval head and pectoral fins. It was twice as long as the longnoses, and as we later determined, most of those fish were four feet long.

Anyway, since Lindsey had laid a new upstream block-net, we collected the rest of the nets, then cruised up and spread them in an area where two small creeks were spilling into the primary stream. We figured that this influx was attracting the shad, and that the shad were attracting gar.

No sooner had the nets been set than the floats started bopping. We pulled out a longnose that was five-foot-three and thirty-five pounds, then another that was four feet long. Then one that was almost four feet, then one that was four-eleven. Reid was recording the lengths in centimeters and I was converting to inches. Most of these fish were thirty pounds plus.

I stated earlier that the thirty-five-pound longnose we captured the year before would've been a state record if caught on rod and reel. At that time, the record for any kind of tackle was a thirty-four-pounder shot by an arrow. A couple of months ago, though, the guy who held that record for twenty years went out and shot a fifty-four-pounder in the Ouachita River, thereby setting a new record.

Photo 9. Torry Cook and his Sixty-Six-Inch, State-Record Longnose.

Photo by Susan Cook.

Still, the ones coming out of this stretch were just as long, and about as big as longnose get. We caught them all afternoon, all the way up to five-foot-four. The largest was probably forty pounds, but it got to the point that we stopped weighing them because they became a nuisance, getting in the way of gator gar.

The first one we caught was five-foot-two, and when we ran a detecting mechanism over its tail, we found it had a PIT tag in it. A "passive interactive transponder" is an extremely small tag that gets injected under the scales with a syringe-looking tool. Over the last few years, Team Gar had shot up forty-seven members of this population and had incorporated a "mark-recapture" formula that can be used to estimate the population based on how many fish get recaught.

I should also mention that we had a new and improved method for hauling these suckers in. Whereas we used to position an awkward stretcher beneath them (which sometimes resulted in losing a big fish or two), we now used the gillnet a fish was tangled in to wrap it up even more, then hoist it into the bow bound up like a Christmas ham.

On this particular sixty-eight-pound alligator gar we decided to attach a floating radio transmitter, which was different than the cigar-shaped kind we usually affix beneath the dorsal fin. Lindsey showed us two models for the new "floaters." One was from Australia, where they're commonly used for tracking sharks, and it looked like an ice cream cone topped with a single scoop. The bulbous part was composed of foam and the tapering end had a cable on it that could be attached directly beneath a dorsal. The "hillbilly model," however, was Lindsey's own invention. Fundamentally, it was the same design, but the bulbous part looked more like a cave rock from *The Flintstones*. It wasn't pretty, but it would do the trick.

The logic of the floating transmitter is threefold. First, it's easier to recover if it comes off the fish. Secondly, since it trails behind the fish, there's less possibility for rubbing to cause an irritation or infection. And thirdly, since it floats, the antenna stays above the water when the gar is on the surface, so it can send out its signal better.

Chris attached that transmitter and we took "Hillbilly" downstream with a load of longnose and let them go below the block-net so they wouldn't get recaught. Then zooming back, we caught even more longnose: five-two, four-ten, three-eleven, four-three. They just kept coming: beautiful fish, healthy fish, one of the biggest old-growth populations in the state, nothing under three-foot-eight. Essentially, this area was ruled by the big boys—and big girls. Not because it was forbidden for the youngsters to tag along, but because, apparently, there hadn't been a successful spawn in decades to supplement these populations of alligator and longnose gar.

This analysis proved to be incorrect, though, because suddenly Lindsey and his crew came back from upstream reporting they'd bagged two gator gar. When they pulled up, there were just two squat humps sticking up from the onboard tub.

Stumpy and Shorty were true to their names. Shorty was three-foot-eight and twenty-four pounds, and Stumpy (who looked like a blimp with fins) was four-feet long and thirty-six pounds. They were the first adolescent alligator gar we'd ever caught in these waters.

Genetically speaking, the gar were back. A new generation was in the mix! Which was something we never expected, but always dreamed of. And because of this revelation, we were stunned. But not so stunned we couldn't wrangle in more gator gar.

This time we handled them differently than we did two years ago. After measuring and weighing them, we immediately placed a wet towel over their eyes to keep them from stressing out. Also, since we were now adept at tagging them and releasing them in a much more timely fashion, we didn't need sedatives or a holding pen. All we needed was to inject a PIT tag if they didn't have one, and then attach a Floy tag, which is an orange plastic-coated wire with a number like FWS118 on it and a phone number to report the fish if caught.

We also took fin clips. Each tissue sample was preserved in a vial of alcohol with a penciled number on a scrap of paper. We recorded where each gator gar was caught, the size of the mesh, and the depth. The water and air temperature had already been logged in.

Another thing I learned while trying to untangle these gar is that there's a simple way to get them to open their mouths, so as to work the mesh free from their teeth. I've seen people bust gar teeth by prying jaws open with rebar, but now there'd be no more of that. As Jaypea explained, all you have to do is tap where the jaws come together, and they'll open up with a big fangy grin. And they did.

All afternoon, we kept catching gar. The longnose became so common that Team Gar started throwing them around with no abandon. We were tossing them from boat to boat, to measure them and release them as fast as possible. It made me cringe whenever someone dropped a fish and it banged around on the bottom of the boat, but gar are tough. And as we've found out, they can take our abuse.

The next alligator gar was five-foot-three. It weighed seventy-five pounds and we called him Big Boy. Then we got two young pups: three-foot-one and three-foot-three, twenty-two and eighteen pounds, respectively. For every full-grown gator gar we were catching, we were catching a juvenile as well—which Lindsey kept saying were from "Tommy's generation." Wildlife agent Tommy Inebnit had tagged ninety-two six-inchers back in 2007 when he'd been a graduate student.

In 2008, Team Gar recaught one of Tommy's generation, which was verified from the PIT tag in it. In a year, that fish had grown twelve inches. Four years later, those gar were between three and four feet. Hence, we were getting a better idea of growth rates for this specific population. It was looking like they were growing about a foot per year. At around five feet long, they start growing wider.

We kept on fishing and got another "recapture," which was five-foot-four and seventy-five pounds. Then, when I was pulling up a net in the hole above the Garhole, I saw a thick gray form emerge from the murk. It was a ninety-pounder, so I had to get some help hauling it in. When we measured that one, it turned out to be five-ten. This one was also a "recap."

It certainly seemed that the big ones were hanging out together: the old school alligator gar and the behemoth longnose. It also appeared that the largest known population of gator gar in Arkansas was running with the oldest known population of longnose in the state. There were literally hundreds in there! It was Garmageddon in there!

At one point, we approached a bobbing float, and I saw something hung up in the mesh. It was Lindsey's homemade floating transmitter.

"Hey," I yelled, "it's Hillbilly!"

Reaching down, I worked the transmitter out. It was a delicate proce-dure, and I was afraid the fish might make a run for it and rip itself up. But for some reason, Hillbilly just hovered there and let me get it free, before sinking straight down and vanishing from sight.

What we learned from this was that the gar we were releasing were sneaking back under the block-nets, because they wanted to be part of the gang. This observation was reconfirmed every time we saw a gator gar hovering under a log or branch. We could see their backs sticking up. Having returned to their pals, somewhat exhausted from being hoisted and restrained and stuck full of tags, they were using brush for balance as they regained their composure.

This made me think of a gator gar named Judas, which Lindsey caught and tagged a few months ago. I asked Lindsey what had happened to that fish.

"Aww crap," Lindsey sighed, then told me how after they released Judas, they saw him pinging right by the Toad Suck Dam a few days later. In fact, it was right under the dam on the upstream side. For the next few days, that's where the signal kept coming from, which was not very encouraging. Lindsey feared that Judas had been dazed from being worked on, so he'd let down his guard and got caught in some sort of suction. Having recently netted a longnose with a broken back and chewed up tail caught in an eddy beneath the dam, I knew those gates could really mess a gar up.

When the Corps of Engineers released water, the signal had disap-peared. Lindsey said he went all the way down Pool 7 and all the way up Pool 8 with headphones on, listening for Judas' signal, but he couldn't find him anywhere.

Most likely, Judas had become a casualty. Which, in turn, made me revise my idea about gars being tough enough to take our abuse. Because really, they're just as tough as us. That is, they're strong when they're strong, but sometimes they're extra sensitive. So if you release a gar, try to do it in a place where it can find some brush for stability. Or better yet, support it until it's strong enough to swim off on its own. There's no better way to bond with a fish.

Meanwhile, the sun was starting to go down, and we were still hauling the longnose in, one right after another: four feet long, five feet long, twenty pounds, thirty pounds—they just kept coming. And I kept recording information—up to the thirty-second longnose. That's when I finally gave up.

But the gator gar refused to give up. We caught one that was five-foot-two, who I figured could've been Junior, a fish we caught two years ago. We didn't have the exact info with us on the fish we'd tagged that season, but judging from the recap number, we could see that this gar was tagged during that run. Junior had been the smallest of that bunch, only measuring five feet long. The thing about Junior is that we'd left him for dead. He was struggling to stay upright—had an air bubble in him we couldn't work out and he couldn't shake.

Whatever the case, we caught another four-foot alligator gar, weighing in at thirty-eight pounds, who somebody named Newbie. Then we caught an eighteen-pound gator gar, who was subsequently named Another Newbie. And as all this was going on, Reid dubbed this stretch of river "the Garburbs."

Then we caught Biggy, who was the most mongo catch of the day. Biggy was six-foot-ten and 166 pounds. This surprised Reid, who thought we'd accounted for all the ancient big ones in this population.

Then we got hit by another surprise, which is something we should've figured out years ago. Basically, the stretcher we were weighing these fish in became heavier after time. For the first fish, the stretcher was dry, but after getting wet and full of slime it took on a few more pounds. Unfortunately, we were now on a system that had been in place for years, so there was no going back and correcting the errors.

But back to the gator gars, which just kept coming in: we recorded another four-footer, then one a few inches longer, and then another a few inches shorter. Until finally the sun went down, and we tallied up a total of fourteen alligator gar, upwards of fifty longnose, and a goofy-looking paddlefish.

Chugging back in the dark, I couldn't help considering Martin T. O'Connell's 2007 study in *Estuaries and Coasts* (vol. 30, no. 4), regarding long-term declines in apex predators in Southeastern Louisiana, where gator gar populations had plummeted 99 percent in fifty years. O'Connell had focused on bull sharks and alligator gar and had stressed that "The removal of apex predators can lead to system effects that may cascade down through multiple trophic levels, fundamentally changing ecosystem structure." O'Connell also stated that "Such alteration of trophic structure can be especially disastrous in highly productive ecosystems." In other words, upsetting the nutritional food chain can result in a loss of species diversity. Half a century ago, this happened in Arkansas when gator gar were removed from the system.

To play the Devil's advocate, it's reasonable to assume that in the grand scope of things the overall health of the planet would only be a micron more stunted if alligator gar populations were to collapse in this region again. Still, millions of microns can add up to the point that they subtract from our "quality of life"—which seems to be the universal measurement for what we believe we're entitled to in an increasingly urban environment that views Nature as a resource to exploit rather than a place to co-exist.

These reflections, however, were cut short when one of the students took out a calculator and entered some numbers. I had no idea that we could calculate the population from data gathered that day, but the mark-recap equation was working for us, and what it revealed was that there were eighty-two alligator gar in this system.

This was considerably less than the hundred I was hoping for. Never-theless, we were encouraged by all the new gar blood in our gar community. This was monumental news for us, for gar, and for fishing in the entire state. Something right was going on in this artery of the Arkansas—where gator gar weed out the carp and drum, where the sweet blue cats are thriving and fat, where the crappie prosper due to all the large gar devouring the jumbo shad. As for the buffalo, they are mammoth in this stream. But the lack of development along these banks, the lack of traffic in this current, and the plant life in the spawning grounds also factor in. And then there's the paddlefish, the sturgeon, the turtles, the birds, the everything that's been cycling here for centuries —especially the top predator: alligator gar.

Meaning that since this is the way our waters used to be, this tributary can be viewed as a model for what can be reclaimed. Because ultimately, my concept of gar making a comeback isn't just some fantasy providing a convenient excuse for a book; it's something real, something actually taking place, something we are directly involved in.

And for me, personally, I'm left with the sense that the small part I play in all this is the most constructive thing I can do. Work like this makes me feel useful, responsible, invested in something larger than myself. And the results are more than just visible; they're right there in front of our faces, and in the data we keep. And what this data demonstrates is that our actions in fishery management, research, outreach, and the creation of new regulations *do* make a difference in preservation and propagation—for the gator gar have returned!

Chapter 5

Gar Rodeo in the Cajun Swamp

Judge Not, Lest Y'all Be Judged Yourself!

"You ain't a activist, are ya?" the manager asked me on the phone. "Because we don't need no PETA getting all up in our grille."

"No, no, no," I replied. "I'm just a gar-writer trying to learn as much as I can. I want to come down and check out your gar-fest, meet the people, see the fish."

He was worried that I might judge their event harshly and get the animal rights folks all up in a lather. For the twenty-sixth year in a row, the Blind River Bar was holding its annual "gar rodeo," a jugfishing tournament in which self-professed coon-asses from all corners of the Cajun swamp converge on the Diversion Canal of Lake Maurepas, Louisiana, for a weekend of good old fashioned redneck revelry and gar-fishing action. Last year, sixty-five boats entered the competition, they brought in thousands of pounds of alligator gar, then had a major gar-feast.

When I was researching gar rodeos (which is what gar-fishing contests are usually called), the Blind River Bar's website stuck out from the rest. They were definitely the most popular one going on in the Deep South or anywhere, and the fact that the bar was only accessible by boat made the prospect even more intriguing. Their gallery of photos featuring buxom barmaids based on the Hooters prototype and their *Coyote-Ugly*-spring-break atmosphere designed for hot young binge-drinkers provided for the promise of a colorful adventure. Since gator gar were at the center of the festivities, I knew I had to investigate.

My main question was whether it was sustainable for those swampy populations of gar to have all these angling partiers descend into the environs of St. Amant, which is pronounced "Sanama," like Panama. There were prizes for the biggest gar, the heaviest load of up to ten, and the largest "trash fish" other than a shark. The only rule for gator gar was that they had to be brought in in edible condition. This, however, didn't mean they had to arrive alive.

There were no limits on alligator gar in the state—for size, for quantity, or time of year for harvesting—and I was concerned about the hit they took from the petro-Pollock that had splattered the Gulf two years ago. So I'd written to Dr. Alysse Ferrara, the gar specialist at Nicholls State in Thibodeaux, asking what she thought. She had replied:

> I don't think jugline fishing is too big of a problem in coastal Louisiana. Our coastal populations appear to be large and from preliminary analyses of two coastal populations . . . the coastal fish are young and fast growing . . . one of the problems I see with juglines is people from non-coastal areas fishing in areas where coastal residents fish . . . When other anglers from outside of the area come down, they generally fish the most easily accessed waters and fish them hard. I think the biggest hazard may be gear that is not retrieved.
>
> If we continue to lose coastal habitats we will lose critical spawning and juvenile habitats. The oil spill probably impacted a small portion of alligator gar habitat in southeastern LA . . . If another spill happens that impacts a larger area further inland during the spawning and juvenile growth periods we could have a big problem. I would expect adult gar to avoid oiled areas but the transfer of contaminants through prey items may occur. Loss of coastal habitats due to erosion, subsidence, and saltwater intrusion

is the biggest danger our coastal populations face. If loss continues or accelerates, a future spill could be devastating.

With this in mind, I rented a condo three miles downstream from the bar, hitched up my *Lümpabout*, and headed down in the August heat, triple digits everywhere. With Robin riding shotgun, and our friends Sharon and Brian coming over from Baton Rouge, we were fully prepared to submerge ourselves in a backwater bacchanalia of seafood abuse and carnival culture with gator gar at the nucleus.

* * *

After making it through a typical torrential afternoon thunderstorm, the first person we met was Captain Keith, rigging up his boat (dubbed *Tea-Bag*) in the parking lot at the condo. Pointing his rum and Coke at a ten-foot-tall monster truck across the lot, he told us that the guy who owned it was Troy Landers' "chooter" from the show *Swamp People*. Like Captain Keith, this local celebrity was getting ready to participate in the rodeo.

It didn't take long for Captain Keith and me to start talking gar. Since I told him I'd written a book on this fish, he was pumping me for as much info as he could get. And I, in turn, was trying to get the skinny on his gear.

He showed me his stuff. He had a hundred floats labeled with his name, all attached to 750-pound-test lines between two and six feet long. On the ends of those he had two-foot stainless steel leaders as thick around as coat hangers, attached to 10/0 J-hooks. He'd be using frozen "pogeys" as bait, which are a type of saltwater shad.

At some point during our conversation, Captain Keith magically replaced his drink with a Coors Light, then told me how his crew would be here in the morning, that they'd register in the afternoon, then gas up and go out, fishing almost all night long.

"I'm only going to drink a twelve-pack," he told me. "Because I have to drive the boat."

Captain Keith had a special spot already picked out in a canal between I-10 and I-12, where he and some buddies had shot some eight-footers with bow and arrows a week before. He even showed me an image on his iPhone of three of them, and they were about as big as alligator gar get. One weighed 175.

Captain Keith warned me to watch out for the law, which he said was all up and down this waterway, pulling boats over all the time. Then he mentioned another creature, which he said we'd see at the Blind River Bar. He referred to this species as "the juiceheads."

I wished him luck and launched my boat. The condo came with a slip on the canal. That's when our friends showed up and we went to the Mexican restaurant for dinner.

Upstream, they were featuring an Elvis impersonator for the opening of the gar party, but I didn't want to run the boat at night. I figured we'd see more in the day, and besides, there was a tiki hut next door.

We went over there, but really, there was nothing tiki about the place, except for a sign telling us it was. When we got inside, it was your typical empty dim-lit dive. Robin ordered a whiskey Coke and Sharon and Brian got some Abitas. I had my heart set on some sort of tropical drink in a long-faced cup, so I got the closest thing I could find: a pineapple-flavored Smirnoff Ice, which I chose to think of as "malt liquor" rather than a wine cooler.

DJ Whatever took the stage. He was an urban whiteboy with a blingful cap and saggy pants, blasting angry gangsta rap. On stage with him were his five or six buddies, dressed the exact same way as him to assert their individual style. They were all about twenty years old, and there was one hot blonde among them, twerking like crazy.

I tried not to stare, but it was too hard not to watch. She was shaking her bootie at 5000 RPMs, the seams of her butt-hugging shorts threatening to burst while the only four people in the audience (us) tried to pretend that a gal the same age as Brian and Sharon's daughter wasn't acting like a stripper. Apparently though, such sexually explicit gyrations were socially acceptable in her community—so who were we to judge her? Nevertheless, I wondered if this scene was some sort of prelude to what we'd be seeing tomorrow.

* * *

Around noon the next day I saw Captain Keith again, drinking off his hangover, so went over and asked him how he hooked up his bait. He replied that the method was to run the leader through the length of the fish so that the hook comes out its mouth. Since gar swallow fish headfirst, this logic made sense.

Then I asked the next question: "How do you get them into the boat?" Like the first question, I had my suspicions, which were immediately confirmed.

"We choot 'em in the head with a .22," he replied. "Can't have 'em thrashing around in the boat."

Captain Keith had never entered this contest before, but he'd lived in this area all his life and had received plenty of info on what other juggers do. And, of course, Cajuns have their own traditional set of rules, one of these being: take a gar out before you bring it in—which I soon found out was the norm in these here parts.

I tried to explain the more humane way of roping them under the pectoral fins, hoisting them over the rail, then sitting on their backs and clamping their tails between your legs to get them under control. I didn't explain my reason for this (which was to keep as many alive as possible), but I thought I'd suggest it. Because why kill a good fish? I mean, wouldn't it make sense to wrap the largest gar in wet towels and let them breathe, then in the end pick the biggest fish and let the others go? But Captain Keith just squinted at me like I might be an activist.

I looked across the lot. There were more monster trucks parked next to the guy from *Swamp People*, some almost a dozen feet high. The sun was blazing away at a blistering pace, and it was getting time to get to the bar, so I loaded the boat and off we went.

The canal was lined with townhomes and condos, boat slips with power-lifts, manicured lawns, screened-in gazebos, all types of yachts, golf carts, and concrete statues of pelicans and black boys fishing. Within these gated communities there were American flags, private patio bars, and multi-million-dollar McMansions. There were already jugs all over the canal, some slowly moving upstream.

Like us—in the junkiest boat from here to Lake Pontchartrain. My spraypainted, bat-finned Bondo-buggy was getting passed by jet skis, party barges, glittering bass boats, and speedboats so long and rocket-shaped that they looked like they'd been designed by NASA. Most of them had names like *My Toy,* or *Lucky Me,* or *We Won the Lottery!* We even saw one called *Bustin' Fun... in the ass.* But again, who was I to judge?

We heard the bar before we saw it. It was pounding out a skull-splitting cocktail of classic rock mixed with rap, and there were boats berthed all around it—most of them the inboard types. But there were also a few jimmy-rigged flat-bottom boats with clunky old Evinrudes looking just as out-of-place as my half-century-old fiberglass craft. These were the fishermen, registering for the rodeo.

Out on the covered deck, we ordered some beers and cheeseburgers and sat down near the registration table. Giant cooling fans were roaring all around us, adding to the chaos of Guns 'n' Roses being blasted at volume 11. It was impossible to talk to the fishermen, so when they'd wander back to their boats, I'd stalk them, introduce myself, then ask them questions.

These guys always replied with a friendly and amused *"Whatchyoo got?"* and it was reassuring to hear that stereotypical Cajun response of *"c'mon,"* which translates various ways. I.e., "No way," "You bet," "Really?" "Get out of town!" et cetera. Having lived for three years in Lafayette and two years in Baton Rouge, I didn't just have a nostalgia for that rich and curvy accent; there's something about this musical tone that basically just calms my nerves. Perhaps it's the sense of humor in it, or the love for life detectable in the syllables that pretty much come singing out.

But whatever the case, I didn't get as much info as I would've liked. This, however, was okay, because I really didn't know what I was after yet. One guy told me he was out to win it, and would stay out all night long. He was an ex-commercial fisherman and was using mullet, as were most of the others. Another guy told me he was out to "kick ass," and when I asked him what he would use for bait, he told me this was "top secret."

Some busty waitresses then posed with an oversized check for $1000, which would go to whomever brought in the biggest gar. Since I brought along my camera, I snapped a picture of this almost painful instance of the old adage "sex sells." And it does.

Photo 10. Lousy Photo by Mark Spitzer.

As the afternoon went on, more locals came on in—particularly the "juiceheads" Captain Keith told me about. There were loads of these pumped-up, oiled-up, bare-chested gel-heads strutting around with razor-wire tattoos snaking around their limbs like vines. They were all just above the drinking age and taking full-advantage of that fact, gesturing and flexing, high-fiving and shooting pool. They came by their daddy's boatloads with coozies in hand and wrap-around shades, shouting and wrasslin' and carrying on.

It was my wife who first brought up the word "Guido," which she can say without being faulted, because she's from some exit in New Jersey where this isn't as much a derogatory label as it is a way to express a common ethnic category in the most accurate and acceptable way the people there are accustomed to. Then Sharon mentioned the reality show *Jersey Shore*, and I realized that this assessment was spot on. These guys were definitely kin of the Situation and Pauly-D. They were laughing at "grenades" and looking to "smoosh." Some were even looking for a fight.

That's when the word came to me: "Guiddeaux." Yep, just adding that French "eaux" that Cajuns love to play with so much seemed totally appropriate, whether those juiceheads had Italian roots or not. Because, whether they realized it or not, that's who they were imitating, so that's what they were.

Still, I was trying to restrain myself from passing judgment on anyone or anything having to do with the research I was doing. This is exactly what the manager was worried I'd do, but more in respect to the portrait I'd paint of how the Blind River Bar treats fish. Since I had tried to reassure him that that I wouldn't get all environmental on his ass, and since he wasn't kicking me out on mine for coming over uninvited, I felt obliged to avoid being critical. The thing is, though, I hadn't even seen any garfishing yet—I'd only seen the clientele.

And the gals weren't any less Jersier. They were falling down and stumbling around and rubbing up against each other as they sang along to drinking songs and boozed to surpass Snooki and JWoww. Robin called them "bikini gals" and pointed out that the fashion was a skimpy top and fake breasts with unzipped too-short short shorts revealing a flash of swimsuit bottoms. By 2:00 p.m. they were puking drunk. After

Robin went to the restroom, she reported that a bunch of them were purging in there in order to continue to eat and drink.

Okay, I figured, there's no way now that I can't *not* show what I see. It would be dishonest and insincere to pass over these realistic details, to pretend I don't see this world as it is. Meaning I couldn't help it. I had to be who I was—so I came up with another word: "Bimbeaux." Which might seem sexist, but my feminist wife and politically correct friend Sharon (the minister who married us) thought the word quite apt.

By three o'clock we'd seen enough, and had definitely heard enough brain-bashing music to go deaf in one ear. I'd talked to enough fishermen too, so it was time to get the hell out. Winding through the beer-pounding crowd, we made it to the *Lümpabout*.

There were now hundreds of people at the Blind River Bar, pushing out their pecs and breasts, silicone bouncing all around. So when my stripy yellow boat pulled out, we immediately became the center of attention. The Giddeauxs and Bimbeauxs were pointing at us. Captain Keith, pulling in to register, was pointing at us. The chooter from *Swamp People* was pointing at us. Even the cops were pointing at us.

There were police floating everywhere, just waiting to hand out their quotas of DUIs. Then hitting their sirens, they pulled me over. Because now I was being judged.

It was stupid of me for assuming that I could just come on over from Arkansas and drop my boat into the bayou. If I'd been better prepared, I would've known that it was a requirement to wear a lifejacket when operating a tiller-driven boat in this state. I also would've known that "throwable floatation devices" are not considered "personal flotation devices" in Louisiana.

So there I was, not up to snuff and getting busted—but not by the Sheriff. I was getting busted by the Department of Wildlife and Fisheries, who carry guns in Louisiana, whereas in Arkansas, I basically work with the same state agency on sampling and tracking gar. These guys, however, were focused on giving out tickets, and they gave me two: one for not having a fire extinguisher on board, and one for not having a lifevest on. They told me I was in violation of four laws total, but they'd only cite me for two (gee thanks, fellas!). Then they gave us some lifejackets and instructed us to return them to their truck later on, which was conveniently parked at the Hill Top Inn, where we were heading for dinner that night.

But wait! Something then happened after the fuzz let us go that has no connection to the experience just reported, except in a strange metaphorical way. Heading downstream, I spotted a belly-up alligator by the shore. We'd seen it earlier, just a little sucker with its tail cut off—a four-footer now down to two feet—which I thought would be cool for a photo op.

Photo 11. Reeking Partial Alligator.

Photo by Robin Becker.

Motoring over, I picked it up and raised it high, waiting for Robin and Brian to focus in. That's when I noticed it totally stunk! It was Rancid!

Putrific! Nauseas, Gaseous, Rank as could be! It was so damn stanky that every molecule in my body was screaming for me to toss it back, but the cameras were still zooming in. Sharon was trying not to hurl, and I was getting to that point myself, but finally they snapped the money shots, I dropped that reeking gator back in, and we tore on back to our dock.

* * *

The weigh-in began at 9:00 a.m. the next morning. Brian and Sharon were sleeping in, but Robin and I, with borrowed lifejackets and a brand-new fire extinguisher, arrived shortly after that. We went over to where the boats were unloading. Only a few had come in, but there was already a pile of skinned gars attracting flies in the sun. Most of these fish were between two and four feet long, and all of them had holes in their heads.

Photo 12. Shot-Up Gator Gars.

Photo 13. Shot-Up Gator Gars.

Photos by Mark Spitzer.

As for the humans, one guy was running an electric crane which weighed fish in the kind of basket that gets lowered from a helicopter to rescue people, and another guy was cleaning them. He stuck their heads into two 2 x 4s forming an acute angle, which held their skulls in place, then chopped their backstraps off with a hatchet, right through their scaly armor. Another guy was making garballs. He'd take the steaks from the butcher, process them in a giant grinder, then mix the meat with instant mashed potatoes, onions, spices, and other stuff.

Photo 14. Gar Cleaning.

Photo 15. Gator Balls.

Photos by Mark Spitzer.

Then there was the Weighmaster, who recorded the weights and checked the fishermen's registrations. I approached him and told him I was documenting the gar-fête, and he immediately granted me access to the cordoned-off area—which was fortunate for me, to be right in the midst of the tournament action. With my camera and notebook, I was able to stand by his side for the next three hours, recording weights, snapping photos, and talking to participants, the media, and those who'd come to enjoy the show.

The clientele now was a lot different than the day before. There were moms and dads and kids galore, and the music had changed as well. Now it was country, and the volume was down to 6 or 7. People could talk, people could laugh—but mostly they cheered whenever a big gar was brought in.

The third boat I saw was loaded with five- and six-footers packed in ice. The guy told me he'd caught over a hundred gator gar, but these were the biggest—and the biggest of his biggest was six-foot-six and weighed 66.3 pounds. They had to weigh this load in two installments. The first basket weighed 342.1 pounds and the second came in at 175.1 for a total of 517.2 pounds. Those fish went into the processing pile.

The fourth boat came in with a smaller load, with a few blue cats to boot, which qualified as "trash fish." The fifth boat, however, had a bigger blue cat: 21.6 pounds. That guy had nine alligator gar, of which three were sixty-pounders with monofilament leaders emerging from their jaws.

The sixth boat wasn't that remarkable, but the seventh boat had 365.6 pounds of gar packed into tarps covered in ice. Then came the seventh boat with 258.6 pounds and a thirty-pound blue upping the trash-fish ante.

After that, a family brought in the world's hugest Ziploc bag, dry-iced with ten big gar—but not enough to place for a prize. A few of those fish were sixty-pounders, which seemed to be the going rate for big ones in this area. I mentioned this to a ZZ-Top-looking old timer standing on the other side of the rope. He was a contest participant, and he just shook his head. "Used to be a lot more a lot bigger," he told me, "but they been fished too much."

Robin was also talking to the locals. At one point I looked over and saw her speaking with a toothless old dude with the thickest Cajun accent I had ever heard in my life. He told her, "Doz gar dare, dey eat twice dare weight in game fish evwy day, dey do!" To which she replied, "Actually, that's not true." And that guy, he just walked away without saying another word.

The tenth boat came in and unloaded a bunch of six-footers. The first basket held 235.4 pounds of gar and the second held 214.1 for a grand total of 449.5 pounds. Their biggest fish weighed 77.2 pounds.

But then the biggest gar-load came in, with no fish less than six feet long. They were lined up like jumbo sardines, packed into the hull. This load topped out at 571.8 pounds. There was whistling all around and the kids went nuts.

Photo 16. Mother Lode.

Photo 17. Deformed Tail.

Photos by Mark Spitzer.

One thing I noticed was that all these gar had a blue sheen. They didn't get red in their bellies like Arkansas gar and most of them had perfect tails. One had a freaky forky deformity, but that was it. The twelfth boat only came in with three fish, but they were three fish that really counted. They had a big fat blue that weighed in at 45.7 pounds, and two seven-foot gator gars, the largest weighing 114.5.

Photo 18. The Big Ones.

Photo by Mark Spitzer.

The next boat was also a contender. Beneath their pile of five-foot gars, I saw a tail so girthy that I figured it was a 200-pounder. But when they finally pulled that fish out, I saw it was the opposite of a big cat. Whereas catfish tend to have mongo heads that taper toward their tails, this gar had a little head and a lot of junk in its trunk. It weighed in at 107.2, and the whole load weighed 463.1.

After that, a flat-bottom boat pulled up with no ice or refrigeration and unloaded 233.3 pounds of alligator gar and three big cats. The Weighmaster directed those spoiled fish to be hauled directly to the ever-growing gar-garbage pile.

Photo 19. Garbage Gar.

Photo 20. Garbage Gar.

Photos by Mark Spitzer.

"Those fish stink," he told me. "All those other gar brought in, you couldn't smell a dang thing. But those guys, they don't care."

With all these big gar coming in, I hadn't stopped to think about waste. It wasn't bothering me that I'd already seen over five hundred gator gar with bullets through their brains, probably because Dr. Ferrara had assured me that this population wasn't at risk. What did bug me, though, was wondering what had happened to the rest of the fish—the fish we didn't see. Like the ninety other gar the third boat reported catching. But talking to some other fishermen, I'd found out that a lot of them brought their biggest ones in and saved the remainder for the commercial market. Or so they said. But those stinky gar, they were wasted. Definitely wasted. Two hundred and thirty-three pounds wasted.

The fifteenth boat then sidled up and unloaded a 29.7-pound cat. I watched another boat unload, and then Robin said it was time to go. Brian and Sharon were waiting for us. So I checked out the four or five boats left in line, saw a few more remarkable gar waiting to get weighed, and also noted an alligator snapping turtle the circumference of a trashcan lid, which would provide some competition in the trashfish category.

On the way back to the condo, Robin asked what the deal was with the other boats. Sixty-six had registered and had paid sixty bucks each, but only a third of those had come in to weigh their fish. I'd asked the Weighmaster about this as well, and he had replied that some people must've got skunked, and that some just knew that what they caught wasn't enough—so that's what I told her.

But then we got another answer to that question when we saw the *SS Tea-Bag* being towed. It was heading downstream, so I pulled up alongside. Captain Keith waved his beer at me and told us that the gators had shredded his lines, and that after spending $200 on fuel, he'd run out of gas.

* * *

We missed the fried gator gar and gar boulette extravaganza, but returned in the afternoon with Sharon and Brian. The music was mega-thumping like usual, and the crowd was a mix of juiceheads, bikini gals, what Brian referred to as "the job creators," and fishermen.

Around four o'clock the garmaids were led out to the deck to pose with the trophies and champs. The MC made a speech about how there should be an award for the drunkest fisherman, and a couple Guiddeauxs pumped their fists in the air and shouted "That's What I'm Talkin' About!"

Looking around, I noticed that everyone was completely white. There were no Latinos or African Americans—which surprised me, but not Sharon, who was a native Louisianan. When I mentioned this to her, she replied that we were right on the edge of Livingston Parish, which is still known for the occasional cross-burning. But again, I wasn't here to judge.

The third-place winners for the biggest load of gar were then announced. David Hunt and Jason Thompson had brought in 463 pounds, so they stepped up and received a giant check for $500. The waitresses grinned and flashed some cleavage, pictures were taken, and then it was time for another prize.

Jerod Galloway was awarded $500 for the largest trash fish, that forty-six-pound cat. He posed with the eye candy, and then Derek Pasternod and his crew did the same for winning second place in overall gar mass: 517 pounds.

Then came the award for most gargantuan gar of all. It was a thousand-dollar check, which Jerod Galloway also won for his 114-pounder. He stepped up with a big ol' smile, the gar gals stuck out their boobs, and jpegs happened.

Photo 21. Big Winner.

Photo by Mark Spitzer.

Jason Snyder and Co. were then awarded the grand prize of $1500 for their 572-pound garload, because each of those six-footers weighed an average of fifty-seven pounds. More flashes, more flesh, more whoops from the crowd.

Then it was over, just like that. The fishermen began pulling out, leaving the Guiddeauxs and Bimbeauxs to return to their rutting rituals in their natural habitat. The cops kept watch, the music got cranked up, and I cornered Jerod Galloway before he could get away.

"Did that 114-pounder give you any trouble?" I asked.

"Besides staying awake since nine o'clock last night," he answered, "*c'mon!*"

And that was all I had—just couldn't think of any further questions. So I went on back to my friends, feeling empty and exhausted. The anti-climactic after-effect was kicking in, so there was only one thing left to do: pay our tab and go back to the condo.

* * *

We took it easy, just motoring along and taking it in: the oyster mushrooms along the shore, which we hadn't noticed before (so stopped and gathered a few), the lily pads, the alligator grass. But there was something else in the canal.

We saw the first one from forty yards away, floating bloated, a rigor-mortised pectoral fin waving to us in the wind. It was only three and a half feet long, with a bullet hole through its head. Then it spoke to me. It said, *Where you from, cher, every gator gar counts. But down here, we got beaucoup to spare!*

Again I looked around. We'd traveled this stretch at least eight times over the last two days, but never saw one dead fish in it. But now, there were alligator gars bobbing under docks and washed up in the weeds like random cans of Bud Light: the excess, the dregs, the casualties of rodeo.

We pulled up to another gar, this one a five-footer. It was a beautiful fish, but after baking in the sun all afternoon, it stunk just as much as that chopped-in-half gator we met. This one also had a bullet through its skull, prompting Robin to ask me how I felt.

But I didn't feel anything. Maybe because I was still digesting the whole enchilada. Maybe because I didn't know. Or maybe I felt I still didn't know enough to know, because I wasn't part of this culture. Because when you're part of a culture—particularly a hunting culture with traditions involving animals—there's a context which those on the outside can never fathom. This was something I had to consider, because maybe there's something about this "sport" which justifies what's left behind.

One thing I knew for sure, though, was that I wasn't passing judgment at that moment. Nope. At that point, I was essentially a journalist, objectively recording what I saw. No opinions, no emotion, no bias whatsoever.

Now, however, my attitude is obvious: I've been playing with stereo-types, so any fool can see you can't trust me. All you can do is take what you've got and make of it what you will, which puts you in the same boat as me. You might be an "activist," you might not. It's your call.

Anyway, as they say, "I'm just saying"—and that's the gar-damned truth.

Chapter 6

Bromancing the Gar

In Pursuit of Trinity River Seven-Footers

When I lit off for Texas in October, I had no idea what the story was supposed to be. To get the research travel grant from my university, I explained that my investigation on "the changing gar-scape on the Trinity River" would examine the effects of the new state laws for alligator gar. Meaning I intended to evaluate the management plans on this fishery now that commercial fishing and bowhunting had been reduced. But as I told my pal Minnow Bucket—who was just as psyched to catch a big gator gar—my real goal was a seven-footer.

We were in my 1999 Jeep Laredo towing my bat-finned runabout. Everything that could've gone wrong already had. That's why we were winding through a rutted farm road in the middle of roadkill-nowhere, detoured by construction and poorly marked roads. The sun was going down, we still needed to buy fishing licenses and groceries, but worst of all, we were in a dry county.

At least I had sponsorship, though. My friend the wildlife writer Catfish Sutton had set me up with Penn Rod and Reels, who had sent two brand new heavy-duty combos: a mongo 330GT bait-caster on a seven-foot Ugly Stik, and a golden 750SSm spinfisher on an equally tough Slammer pole designed for hauling deep-sea dino-fish up from the depths of hell. Both of these were equipped with 100-pound woven test. I also had support from Daiichi Hooks and Tackle, who had sent hundreds of bucks' worth of gear, mostly gynormous circle hooks.

So it wasn't just Minnow Bucket and Hollywood gone fishing: It was us plus expectations from my university and two corporate sugar daddies who had invested in the idea of another gar book, which didn't even have a publisher yet. Whatever the case, the sun was setting, the pressure was on, and there was no booze in sight.

Still, we made it to the Walmart in Athens and got our licenses, two steaks, some cans of chili, potatoes to fry, Gator Ade, etc., then shot on over to Caney City. The canned beer there was a huge disappointment, but they had some Dogfish Head IPAs. I bought a six-pack and some gin and tonic, and Minnow Bucket got a case of Heineken, a bottle of tequila, and five bags of ice.

It was dusk when we hit the river, took off upstream, and made it to the sandbar I'd discovered three years back. It was on a bend of the Trinity and the twenty-five-foot hole in front of it was roiling with six- and seven-foot gar. That's what I'd seen when I came up to fish with Jeremy Wade, then went off on my own. Now, however, nothing was rising, nothing was rolling, and the sandbar I'd been dreaming about camping on was covered by a foot of *mudge* (a cross between sludge and mud) that sucked our shoes right off our feet, got all over everything, and promised to be the bane of our existence for the next three gar-mucking days.

Nevertheless, we started in on the beer.

* * *

Before taking off, I had emailed David Buckmeier at Texas Parks & Wildlife asking for his perspective on how the gator gar populations in the Trinity were faring since the new harvesting regulations had been passed in 2009. Buckmeier was in charge of all things gar in Texas, so

I figured he'd know—and he did, replying that "the public's opinion of alligator gar seems to be changing toward conservation."

He also attached a study titled "Alligator Gar Movement and Macro-habitat Use in the Lower Trinity River, Texas." Granted, we were on the upper spectrum of that river, but this report by Buckmeier, Nathan G. Smith, and Daniel J. Daugherty was highly relevant to where we going and what we were doing.

To break it down, the study explained how the researchers had used acoustic telemetry to study effects of flow on alligator gar. In essence, they had documented what Minnow Bucket and I suspected was the case in Arkansas: that alligator gar "moved into tributaries and inundated floodplains during large flood pulses"—which accounts for why we'd hardly seen any full-grown adults in the Garhole over the last few drought-stricken winters.

As for migration, the study noted that "research to date . . . suggests that although alligator gar have the potential for long distance migrations, linear home ranges might be relatively small"—like less than twenty-five miles. This was what Ed Kluender told me regarding the gar he tracked in Arkansas, which never ventured more than fifteen miles from their wintering hole.

The fish for Buckmeier's study "were collected using rod-and-reel, jug lines . . . large-mesh . . . [and] heavy-duty, multifilament gillnets." A cordless drill was then used to attach ultrasonic transmitters with fourteen months of battery life that corresponded to eight submersible underwater receivers (SURs) that detected fish and recorded data. They had some other SUR stations out there, but they got vandalized.

One interesting finding was that "Detections of tagged alligator gar at SUR stations near deep pools where fish were observed . . . tended to be highest during the night." Minnow Bucket and I had been fishing for gator gar all summer, and had noticed an increase in hits an hour after the sun went down. Most previous studies note that alligator gar feed primarily in the morning, so I'd thought it was unusual to experience such vigorous nocturnal activity. Buckmeier's study, however, reaffirmed what Minnow Bucket and I had speculated was the case, sitting out there, drinking beers. Or, to put it in other terms, researching gar "Toad Suck style."

* * *

I may be obsessed by gar, but there ain't nobody more gung ho for gar than Minnow Bucket. He's also a better fisherman than me, which often results in him getting gar and me getting squat. Since I'm supposed to be the "gar guy," this caused me some consternation at the beginning of the summer, but after a few months I made my peace with the fact that he was hooking them and I wasn't.

Getting to this point involved two major steps. The first was redefining my objective, which was to get an Arkansas gator gar before the summer was over. The main reason I felt the pull to do this was because I felt I owed it to my sponsors to bag one with the gear they supplied. But stuff came up and I wasn't able to get out on the river as much as I wanted. Rebuilding my rotted-out transom took weeks, then weeks more finding and sealing the microscopic leaks. It was also the hottest summer on record, with the heat index topping 110 for two months straight. So in August I decided to give myself a break and resign myself to the fact that I caught my dream gar a few years back, so didn't have to prove anything to anyone.

Besides, there's no shame in assisting. That's what I did when Minnow Bucket hooked a forty-three-incher on a sandbar above Cadron Creek. When he horsed it in, I ran out and blocked its run back into the river. That gar then decided to shoot along the shore in just a few inches of water, but I raced alongside it, and when it turned to try and ditch me, I dove down in front of it, scooped it up, hugged it to my chest, then carried it up onto the beach. It weighed twenty-one pounds, a beautiful healthy juvenile. Catch and release, of course.

My point being: Assisting Minnow Bucket with a gator gar for our own research wasn't much different than assisting the US Fish & Wildlife Service in sampling and telemetry. Because that's who I am: I'm an assister, whose assistance comes in many forms, the most important form being typing these words right here, right now, to entertain and educate for the purpose of propagating gator gar. And it works. My research got noticed by Animal Planet, NatGeo followed, the book came out, and word got out that gar don't maim and kill, that they're vital to ecosystem stability, and the big ones need our help. And because of this increased awareness, gator gar are better off. That's what I keep telling myself—to the point that I actually believe myself. For the most part.

The second rationale for not getting bummed out at the fact that Minnow Bucket catches more gar than me is that I've learned a lot from him. The main thing I've learned is that what works on the Trinity doesn't work on the Arkansas. For years I tried giant hunks of smallmouth buffalo and whole drum on heavy-duty rods and reels with shark-sized treble hooks but hardly ever got a bite in my home state. Minnow Bucket, on the other hand, would catch sunfish to use as live bait. He'd cast them out on lighter weight catfish rods with smaller semi-circle hooks (like size 6/0 to 12/0 shiner hooks and wide gap octopus hooks), and he got a maddening number of runs over the summer. Sitting there in our fishing chairs, we'd be just about to give up when suddenly we'd hear a *clack-clack-clack* arise from a bait-clicker. It was always one of Minnow Bucket's rods, and nine out of ten times we'd wait it out to the point that the gar would either drop the bait or it'd get off right before he got it to shore. But sometimes he'd bring one in: a longnose, a shortnose, and once in a while, an economy-sized gator gar.

At first I was pissed—that Minnow Bucket was getting all this action and I wasn't. Sometimes I was even secretly glad when a gar dropped his bait and he'd throw his hands over his face and howl his lament to the stars above. It didn't make me feel the way I wanted to feel, though, to see myself snickering at my buddy's misfortune, so I knew I had to change my attitude. The way I did this was through the if-you-can't-beat-'em-join-'em approach: I decided to use bream like him. I even bought a similar Abu Garcia Ambassadeur 6600 RCX reel and equipped it with fifty-pound braided line. But did I get more hits after that? Not really—but at least I felt like less of a jerk.

Minnow Bucket and I often discussed the merits of sunfish vs. members of the minnow family. For one thing, any type of gar would bite on a sunny, which expanded our probabilities of catching something. Minnow Bucket's theory was that there's a lot less competition in the Arkansas River, compared to the Trinity. In the Arkansas, the big ones are less concentrated, so therefore less desperate for resources. Plus, there's a lot more sunfish per capita in the Arkansas, which is pretty murky, but not as muddy as the Trinity. Hence, you get a higher concentration of carp, buffalo, shad, suckers, and drum in Texas, because these fish—unlike sunnies—don't require gravel beds to spawn. The upshot of his theory being: predators eat what's most plentiful in their environments—and the experts agree with this.

I had a slightly different theory: That when the main populations of Arkansas gator gar got wiped out in the fifties, they were basically overfished out of the rivers with rough fish as bait. So maybe the leftover alligator gars, the ones that didn't get caught in the rivers, maybe they were used to eating other fish. And maybe their spawn, and the generations that followed, had a preference for bream.

My wife argued that this was ridiculous; that a species can't just change its traits in a few generations. But I countered that it wasn't just a few generations. I suggested that millions of years of evolution in the region might've developed a taste for sunfish in certain strains of gator gar, especially those in clearer water, where sunfish tend to swim. Like maybe after the big river gar got fished out, the lake-locked gar and those in the deltas of mountain streams got back into the rivers and brought their numbers up.

That's all speculation, though, and it might be that both Minnow Bucket and I are right. Perhaps both of our theories worked together with other factors to make sunfish preferable for gar in the Arkansas River. But one thing's for sure: Minnow Bucket caught two good juveniles in a year, and neither of us have caught anything on big old slabs of meat.

Still, there's another thing I learned from Minnow Bucket: the utility of the Carolina rig—a concept I never knew of before. Basically, there's a weight that sits on the bottom with a hole going through it, and the fishing line goes through that hole. So when a fish takes line out, it can run for hundreds of yards with no resistance whatsoever.

Anyway, I still feel a bit of competition when it comes to garfishing with Minnow Bucket, but for the most part I feel a kinship in what we quest: prehistoric monster-fish. That's what we've been fishing for, and that's what we're on the Trinity for, linked through the Brotherhood of the Gar. And as long as he keeps reeling 'em in, I'll be more than glad to assist.

* * *

Or so I thought.

After setting up camp, we broke out the bait: seventy-five pounds of carp, buffalo, and drum on ice, compliments of Fishman. Then we were in the dark, kicking back on canvas chairs, digesting two greasy steaks grilled on the campfire. And, of course, we were enjoying a libation or two. Maybe three. Maybe four.

By the time maybe five or maybe six came around, the half-moon had lit up the hole with a ghostly luminescence. The mosquitoes weren't too bad, and Minnow Bucket had his "Mexican wormline" in the water. I don't know why we called it that, but it's a light-weight pole for catching bait, with a one-ounce weight on the end and four small hooks a yard apart. Minnow Bucket sometimes describes this set-up as "a trotline on a stick."

Anyhow, we kicked back and started talking smack. Like usual, I was getting on his case for bringing thrice as many poles as me and taking up space in the boat. I had two rods stuck in the sand (one with a worm, one with a two-pound chunk of fish) and he was maintaining six, so he kept running back and forth to reel in and cast out again. Eventually, he caught a small blue cat on the Mexican wormline and threw that out as bait.

Then, sometime around maybe seven or maybe eight drinks, I was down to one pole and he was only using three when we heard a bait-clicker start to click. It wasn't a long, steady, heading-out-to-deep-water click; it was more like a *clickety-clickety-pause-pause* click—which meant a catfish.

It didn't take Minnow Bucket long to haul it in, all eely ribbed and kicking up a fuss. Too lazy to rise from my chair, I watched him pull it up on the sand, then stagger off in search of a stringer.

"Looks like a six-pound flatty," Minnow Bucket slurred, meaning a flathead cat.

The next day, though, we found out it was a fourteen-pound blue with an ugly white splotch on its head. It was also a cannibal, since he'd caught it on that smaller cat.

After that, Minnow Bucket fell down once or twice (a tradition which, according to him, signals a successful night), then trudged off to his tent while I passed out in my chair, wrapped up in my sleeping bag, pole extending from my lap.

From that point on, the night was a blur. Every twenty minutes or so one of our clickers began clicking. If I wasn't jumping up and groggily releasing line, then waiting for a suspected gar to make its second run, Minnow Bucket was lurching from his tent and doing the same. We couldn't tell if half these runs were the wonky currents of the hole or actual fish messing with us, but they felt like gar—definitely.

Minnow Bucket had a few small pieces of buffalo out and I had half a gou (my preferred word for freshwater drum). Whereas I was betting on the big boys by using large tackle and big bait, Minnow Bucket was going for anything from a medium cat on up—so of course he kept getting more hits than me.

It got to the point that I kept waking every time his bait-clicker clicked. But those fish, they just kept dropping the bait. So around three in the morning, I took my pole up to my tent and crashed out. Still, every half-hour, Minnow Bucket would blast from his flaps and battle a phantom fish. And all the while he was doing this, he would narrate what was going on, blow by blow.

For example: "It's a runner!" he'd yell. "It's taking it, it's taking it! Yeah, it's definitely on! It's spooling me, it's spooling me! Guess I better set it. Okay, okay, I'm gonna set it... here goes!" A second would pass, then: "DAMN! That badboy got off! What a rip off! You Mofo! Why you gotta play me like that? I demand my money back!"

So after his eighth or ninth fight of the night, I wasn't about to haul my sorry butt out of bed. But the thing was, there came a point when he actually had a fish on.

"Hollywood!" he yelled. "Are you sleeping?"

"I'm trying to!" I snapped.

"I could really use some help!"

I decided not to reply. Instead, I just lay there listening to the splashathon going on.

"It's a big bastard!" Minnow Bucket shouted. "Come to Poppa!"

But that big bastard didn't want to come to Poppa. It kept on slapping and taking out line, his drag screeching urgently.

Finally, I got up—but not to help my pal. I got up to see the fish.

The sandbar was shining with a weird fluorescence when I crawled out of the tent. He had the gar along the sand, parallel to the shore, and I could see it in the moonglow: a serpentine four-footer rolling in a wave of its own creation—like it was wrapped in a skin of water. I even saw it smiling inside that crystalline tube. It was a strange, sudden, eerie sight, and in less than a second—SNAP!

"It Broke The Fuggin' Hook!" Minnow Bucket cried.

I stood there for a second, no reply, then stumbled back into the tent as the swearing rose into the sky. Something about "Knob-Slobbing!" and "Dob-Gobbling!" but those were just the adjectives. Something about the hugest fish he ever *could've* landed on rod and reel. Something about "Thanks for the help, bro!"

Right before breakfast, though, Minnow Bucket got another hit, and we didn't even hear it make a run. I simply looked over from the fire, saw Minnow Bucket bringing it in, and a minute later a small silver gator gar came splashing in.

This time I jumped up for the assist, leapt into the Trinity, pinned its head to the sand, got both hands around its girth and carried it up on shore. It was a three-footer, weighing thirteen pounds. And even though I was glad for my friend, the score was clear to both of us: Minnow Bucket 3, Hollywood 0.

Photo 22. Minnow Bucket and Juvenile Gator Gar, Trinity River, TX.

Photo by Mark Spitzer.

All morning long, they were out in that hole—six-footers, seven-footers, maybe even eight-footers, rolling up a storm. All morning long, they just kept porpoising as Buckmeier's words replayed in my head: "the public's opinion of alligator gar seems to be changing toward conservation."

This made me think of Bubba, the guide Jeremy Wade had hired for the "Alligator Gar" episode of *River Monsters*. When I first met Bubba, he actually threatened me, telling me he'd throw me out of his truck if I got all PETA on his ass. Or something like that.

Bubba ran this bowhunting outfit called Garzilla, in which he'd take trophy bowhunters out on his airboat to shoot the big ones. Bubba had claimed there were enough mongo gar in the system to spare, and he let his disdain be known for the less-than-masculine practice of fishing with rod and reel. Still, he condescended to try it out and actually tossed some bait in the water. When Jeremy hooked that six-foot-eight lunker that they didn't even show on the show, then landed it, then let it go, Bubba wasn't so skeptical.

A year later, Bubba changed the focus of his business. He no longer took bowhunters out, he only fished with rod and reel, and he always practiced catch and release. Why? Because he knew the supply wasn't as plentiful as he had claimed, and that if he planned to continue making a living off gar, then he had to do his part to preserve them.

A year after that, I saw an article on GoFishn.com entitled "Texas Drought on Alligator Gar," with a photo of what looked like the cracked terrain of the Sahara Desert with a mud-hole in it. In that mud-hole, there were forty or fifty log-looking gator gars waiting to die in the sludgy soup. That picture had been taken by Bubba, and he'd written the caption under it.

To paraphrase, he explained that this place used to be a lake, but because of a month of 110-degree weather, it had shrunk into a shallow swill-hole. He also wrote about how that water was so dang mucky that the gar couldn't even breathe it, so kept raising their heads for gulps of air. Bubba contacted Texas Parks & Wildlife and they went and rescued those fish.

Good for you, Bubba! But better yet, good for the gar.

* * *

By one o'clock, it was ninety degrees and we knew they wouldn't be biting until after dark. Still, that didn't stop us. We went upstream and tried some spots, getting roasted by the sun. I had two beers and Minnow Bucket had ten or twelve. When late-afternoon rolled around, we were back at camp, baking in the heat index, which was somewhere around 100 degrees.

Both of us then took a swim with alligator gars as large as us chilling out in the cool of the hole. After that, I tooled on over to the other side and crashed out in the shade, fishing poles propped in my lap. Minnow Bucket did the same in a camp chair on the sandbar.

I napped for twenty minutes but awoke when Minnow Bucket yelled, "Hey Hollywood, check out the snake!"

It was swimming right between us, downstream, with no concern for the apex predators hunkering right under it. Probably because it was a rattlesnake: bright gold, four feet long, with a checkery pattern on its back. That snake rode high, more out of the water than your standard copperhead or moccasin, just winding along like it didn't have a care in the world. In fact, it made a beeline for Minnow Bucket, then pulled up on the shore ten yards downstream from him.

"It's a timber rattler!" Minnow Bucket shouted.

I figured it was attracted to the deadfall behind the sandbar, which was no doubt filled with mice and voles and random rodii. And that rattlesnake, it didn't give a crap about Minnow Bucket being there. It just flicked its tongue and stared at him while he broke out his camera. Then, when Minnow Bucket went running toward the snake, it didn't just stand its ground, it charged straight toward Minnow Bucket, letting him know whose turf this was.

Minnow Bucket stopped short and so did the snake. They faced off, just a few yards from each other, and that rattler didn't even coil up or shake its tail when Minnow Bucket got all up in its grille. He shot a bunch of pictures of it, and then it slithered on.

Photo 23. Trinity River Timber Rattler.

Photo by Ben (Minnow Bucket) Damgaard.

Later that night, we had another wildlife encounter, this time with wild pigs. We heard them in the dark, splashing and squealing and snorting around, but we couldn't see a thing. Then they came clamoring right over to our camp, so Minnow Bucket grabbed his gun. I didn't know he'd brought one along, but there he was waving it in the air: a chromy, blunt .45.

I was shining my spotlight on the bank, where I saw what looked like the silhouette of an oval-shaped basketball scurrying toward a pile of brush. I figured it was a piglet, but couldn't tell for sure.

As for fishing, we didn't catch jack that night. There were a couple runs, but that was it. I passed out in the camp chair again, and Minnow Bucket repeated the antics of the night before—jumping up and battling gar, in and out of his tent all night. I even fought a few myself. Still, there were a lot fewer hits.

By morning, it was overcast and colder out. At one point, something large pulled down a Gator Ade bottle I was using as a bobber to suspend a chunk of drum. Then three minutes later, it popped up fifty yards away in the eddy, exploding with a plasticky "KRACK!" But overall, the gar were less active, and we had to put our jackets on.

Minnow Bucket broke out his smart phone and saw a storm heading in. Then he went to weather.com. "It's going to be thirty degrees tonight," he told me, which neither of us were prepared for. When we had left Arkansas, the weather report said it would be sunny and bright for three days straight, never dipping below fifty degrees.

We were therefore forced to skedaddle.

After breaking camp, we headed downstream, me wondering, *Is this it? Is this all we get? No more gar, no more nothing?*

It didn't make sense. Where was the story? I mean, my university paid for our travel, and Penn and Daiichi had thrown down as well. And what did I have to show for it? A tale of two drunken brochachos dicking around in the mud? Sure, we caught one small gar, but other than that we were coming back with nothing more than a close encounter with a snake and some razorbacks.

Then, as we approached the launch, we heard the country music twanging from above. There were two tents under the bridge, a generator was chugging away, and some sleeveless stereotypes on four-wheelers were giving my boat the thumbs up. It looked like a camp for homeless people, but they weren't.

They were just some beer-swigging rednecks fishing from the concrete rubble. And when we pulled up, we saw what they'd left on display: a freshly dead gator gar, only two feet long.

"I can't believe it," I told Minnow Bucket. "I just don't understand it—why people are so proud to kill these fish, then let them stink up the launch for everyone else! I mean, what's the deal with this stupid tradition?"

Minnow Bucket just shook his head, and we got out.

Heading to my Jeep, I considered going over to those yahoos and telling them what the deal was. But that, of course, wasn't the story. The story—and the story that found me—was that I grabbed that beautiful baby gar and packed it in ice, drove it on home, and took it to the taxidermist—

to skin and stuff and mount for my wall. Because I wasn't going to let it go to waste.

So that's what I brought back. And a few months later, I put it on my wall, right next to my TV set, frozen in mid-leap, gills flaring, fangs agleam—a symbol of what could've been, and what's still possible, in Texas and beyond.

Photo 24. Trinity River Gator Gar.

Photo by Mark Spitzer.

Chapter 7

After the Florida Gar

Navigating the Glades of "Deep Connectivity"

On the flight to Miami I dug into the information I'd collected on *Lepisosteus platyrhincus*, or as it's commonly known, the Florida gar. Out of the five North American species, this is the only kind of gar I hadn't yet caught. If it wasn't the rarest of its kind (being endemic to Florida, Georgia, and a corner of South Carolina), it was generally considered an economy-sized gar that hardly ever exceeded three feet. The world record was a ten pounder, but most of them weigh just a few pounds. Fishing-worldrecords.com, however, lists a suspect twenty-one-pound Florida gar caught in Boca Raton in 1981, and a thirty-two-pounder caught in unknown waters by unknown methods in an unknown year. An exceptionally long four-foot-four Florida gar was also documented by various credible sources.

Still, from the lack of literature regarding this fish, it was pretty much evident that not a lot was known about it. Biologywise, there were only a few peer-reviewed studies, mostly having to do with movement patterns, immune systems, reproduction, and the amount of metals in tissue samples. Their closest relative was the spotted gar, which Florida gar differ from in respect to number of caudal rays. The Florida gar also lacks scales beneath the throat. Hybrids between Florida gar and spotted gar had been documented, but for the most part, their territories didn't overlap.

What struck me most about the Florida gar, though, was the amount of overall uncertainty about it. For example, there was definitely some confusion as to when their spawning seasons happened. Some sources said it was between May and July, but others said February through March, whereas others noted that they sometimes spawned as late as October.

Another example of this uncertainty was reflected in some print-outs I had from the University of Michigan's Museum of Zoology website, which claimed that "sex ratios initially lean more toward males but females live longer and grow to be much larger." But as we're discovering with other gar species, particularly alligator gar, the myth about females always being larger than males is a bunch of baloney. Sure, there might be more larger females out there, but the Museum of Zoology's insistence on information that could be outdated didn't make me feel like I could trust the bulk of published information regarding Florida gar. This website also repeated misinformation from the thirties that Florida gar "eat a large amount of fish in a short period of time, devastating populations of other fish." Additionally, there was a breakdown in logic in the claim that Florida gar "are consumed in considerable numbers in Louisiana, and are supposed to be a relatively good substitute for lobster," because Florida gar don't exist in Louisiana. Nevertheless, the idea that gar meat could be used as a type of faux lobster was intriguing.

As usual, the aquarium trade and their enthusiasts were also spreading disinformation. Whereas I found some useful tips on how to use a brief starvation period to wean Florida gar off live food and on to frozen and freeze-dried diets, I also found a lot of chat-room chatter about how a single Florida gar needs at least a 180-gallon tank to live a healthy existence. The Florida Museum of Natural History's profile on the Florida gar echoed this sentiment, noting that "it is unsuitable for home aquariums due to its large size and large tank requirements." This info, however, is total hogwash, because I have a seventy-five gallon tank in my living room right now with a one-foot spotted gar, a two-foot hybrid gar, and a nine-inch bowfin, and all of them are thriving. In fact, I've had that hybrid for ten years and the spotted for almost eight.

I found, as well, that there's further reason to question the industry of exotic fish, which has taken a special interest in the Florida gar. If you look on eBay, or Google the words "Florida gar" and "aquarium," you'll find that the Florida gar is a popular fish for enthusiasts, which can be ordered through PayPal and shipped with ease. And the reason for their success in the tank, I think, is that contrary to the claim that Florida gar are big and need space, they're actually quite small for gar, so they're easy to feed and keep in small tanks. That's why PetSmart has been known to sell them, along with specialty distributors like Rainforest Farms International, who brand them as "banded gar" and sell three- to four-inchers for $19.95 a piece.

The main problem that comes from Florida gar commerce is exactly what happened with snakeheads: People eventually let them go in places they're not from. The US Geological Survey from October 2012 on nonindigenous aquatic species reports a Florida gar being caught in a pond on the outskirts of Salem, Oregon, in 1999, which was promptly disposed of by the Oregon State University Ichthyology Collection. Why? So we don't end up like Canada, where the spotted gar is endangered.

As the Canadian government states in a 2012 Fisheries and Oceans study called "Recovery Strategy for the Spotted Gar (*Lepisosteus oculatus*) in Canada," "The exotic Florida Gar has been collected in the Great Lakes basin (likely the result of aquaria releases). This related species could represent an additional threat to the Spotted Gar, either through hybridization or competition, if the species becomes established." And they have. Florida gar now have breeding populations in Canada. Thus, Canada is seriously assessing the situation to determine the "level of threat to the Spotted Gar," since, according to the recovery strategy study,

"The presence of Florida Gar may exclude Spotted Gar from preferred habitat and cause increased competition for prey."

A 1969 study by Claire M. Bradshaw, L. William Clem, and M. Michael Sigel should also be mentioned. "Immunologic and Immunochemical Studies on the Gar, *Lepisosteus Platyrhincus*" found that Florida gar were capable of producing antibodies to salmonella and diphtheria, but to what extent it's not clear. This study has not been followed up on, but perhaps if it were we might discover some major medical value to be derived from gar.

Thus, it looked like the fish I was investigating held a lot of possibilities. If anything, not much was known about the science of this particular species, there was still a lot of information to question, and there were millions of them down there just waiting for me to cast into their midst and cast all this research aside by finding out for myself what their real deal is.

* * *

I had hired a fishing guide named Jim Dussias, whom I found on an Internet fishing forum talking about the "world-class" Florida gar he'd encountered in the canals of Miami. Jim was also a hunting guide and knew the area. When our conversation moved from the email to the phone, he told me he caught Florida gar all the time and we could get them on flies. More importantly, he didn't laugh at me for thinking of gar as sport fish, and was actually enthusiastic to forego tarpon and snook in favor of the lowly gar.

In the past, I'd been critical of fly-fishing. As a Midwestern fisherman transplanted to the South, I thought in terms of lures and bait, and I liked to joke that trout were trashfish and that the way I caught them was to go upstream with a couple jugs of radiator coolant, pour them in, then go downstream and gather them up.

My real reason, though, for looking down on trout, came from my biased past. Remaining true to my rebellious spirit, I looked at fly-fishing as a bourgeois activity that took hundreds of dollars of specialized equipment and lacked the grit of getting dirty in the name of fish. This supposed "sport" just seemed too clean for a gar-grubbing mudster like me, and I wasn't impressed by the grace and precision it took to loopity-loop a fancy fly into a pristine mountain stream. Give me a worm, was my motto. Or a stinking chicken liver!

Still, I knew that my attitude toward trout was just bluster—and an adolescent bluster at that. I mean, here I am forty-seven years old and still holding to a stereotype that was limiting my growth as an angler and a human being. So after an extensive conversation with my idealistic former self, I figured it was time to give trout a chance—in order to prepare for Florida gar.

For my birthday I surprised my wife by asking for a fly rod. One hundred and forty dollars later, plus sixty more for gear, I was decked out in waders and standing in the middle of the Little Red River. I'd had a casting lesson filled with a lot of dos and don'ts so had decided to set off on my own and figure it out through trial and error. Jim had told me that if I could get the fly in the water, I could catch a Florida gar, and that's what drove me on.

So there I was, the exact kind of fisherman I used to rail against, developing my own guerilla technique. Not only that, I was equipped with $200 of specialized equipment, when I should've spent ten more bucks on a net. Because when a trout nabbed my fly, I was suddenly bringing it in, and then it was in my hands, too slippery to grip. The shore being too far away, I found myself juggling a foot-long, technicolored rainbow trout. Predictably, it broke the line, leaving me more hooked than it had ever been. The fever was on! What a blast! I had to have more trout!

It wasn't that hard ignoring my former self scoffing at me for selling out, but it was hard getting another trout to bite. I kept at it, though, even getting up early on school days and driving up into the foothills, only to get skunked again. Still, every failure happened in a sparkling mountain stream and led to more respect for the skill it took to get that fly where I wanted it to be.

I was also pumped up on stories I'd heard from a former student about fly-fishing for needlenose gar in the Ozarks. It wasn't a very popular approach, but from the articles I'd read, I knew there were fanatics out there who claimed fly-fishing for gar to be the ultimate extreme angling experience. I'd also watched a few videos on YouTube, so was looking forward to seeing if the challenge was worth the effort. In short, it was now time to make my own story, and I was raring to go.

* * *

The plan was that we'd meet Jim the day after Thanksgiving—"we" being my nephew Tommy and me—while our wives went shopping. But I shouldn't say "Tommy," because Tommy is now known as Tom. He's not a kid anymore. He's a six-foot-four professional computer guy, and, like me, new to the realm of fly-fishing—especially for gar.

We got to the gas station at 8:00 a.m. on the upper edge of the Everglades. Jim was waiting for us in the parking lot, and as the three of us shook hands, it was obvious to me that Jim—who was a part-time fishing guide and a full-time art teacher—was a guy I could relate to.

We hopped in his van and he drove for a mile, then stopped by a chainlink fence along a canal. Jim said he'd come here the day before and had seen hundreds of gar just waiting for us.

Basically, this place consisted of two spots: The canal running parallel to the road, and a more concealed "honey hole" on the other side of the levee. That's where Jim took us first. Looking into the crystalline water, we could see all sorts of fishy silhouettes. It was the perfect spot to try some casts.

I immediately caught a peacock bass, and so did Tom. They were a brilliant, metallic, olivine green with bright crimson spots on their tails. So we cast some more and caught a few largemouth, then a couple monster oscars. And they just kept biting, not spooked by us at all, always shooting for our flies. We saw every single one of them strike and could've stayed there all day, but then it was time to move on.

The canal was now boiling with Florida gar. They were surfacing every twenty feet, gulping air, then diving back down. Most of them were about two feet long.

I took a few casts with my fly and nothing hit. Still, I had a secret weapon I was eager to try out. I'd made what's called a "rope fly" and it was attached to a spinning reel on a telescoping rod I'd brought along. Having read about these lures for years, which are supposedly highly effective for gar, YouTube had finally shown me the way. I took a white nylon rope, cut a few inches off, singed one end with a lighter, then tied a regular old knot right below that. Unraveling the filaments, I took a jig head and stuck the barb into the melty spot.

This method, of course, has been used for centuries by various cultures, the logic being that when a gar hits, its teeth get caught in the fibers. Some primitive tribes, even to this day, use spider webs trailing from kites to hook gar-like fish the same way.

I chucked that sucker into the canal and began reeling in. The fly was just a few inches below the surface and a stripy young gar started to follow. As gar tend to do, it sidled up right next to the rope-fly, eyeing it peripherally. Then it snapped, but failed to connect.

Almost every time I cast that lure out, a gar would get up next to it, then follow it toward shore. Nine out ten times they didn't bite, but once in a while they'd take a snap—but no cigar. Tom was doing about the same, using one of Jim's specially designed gar-flies (which was the same scintillating green as the peacock bass and trimmed with golden thread).

Jim gave us some pointers on how to hold our wrists and play the fish if they should bite, but for the most part, the fish were the ones playing us. They were interested in following whatever we put out there, but they weren't really committed enough to take more than an occasional lackluster chomp.

After an hour, Jim suggested a basic floating Rapala, so I tied that on, and that's what did the trick. Wiggling like an actual fish, it attracted a two-foot gar, and POW! it was on.

I didn't give it a chance to fight, just yanked it up on shore and grabbed it. That gar, however, was the trickiest fish I ever tried to grip. It was strong and quick and snapping like a whip as I fumbled up the bank trying to contain its thrash.

It took the three of us to hold it down and weasel the treble out of its mouth. All three hooks were caught in its pallet, so as Tom and I each held a jaw with a pliers, Jim worked the barbs out. Meanwhile, I admired its slightly green sheen and dark spotty stripes. Being my first Florida gar, I was mesmerized.

We tried to take some pictures, but that muscly, determined gar wouldn't hold still. It was leaping from my grasp like a crazed bowfin. Nevertheless, I kept catching that spastic, spunky, scrappy fish, and it kept slapping free, therefore proving to me that Florida gar are worthy of being classified as "game fish"—even though they're not.

Photo 25. Gar Juggling.

Photo by Tom Becker.

Photo 26. The Money Shot.

Photo by Jim Dussias.

After letting it go, we decided to violate Jim's cardinal rule of fishing: "If there's fish in a spot, don't leave it in search of another one." But since those Florida gar were on to us, we cruised into the Everglades along the Tamiani Waterway, which Jim said was blasted from the earth for hundreds of miles after the turn of the century, thereby creating a highly unique fishery.

I'd been reading *The Tarball Chronicles* by my friend David Gessner. It's an eco-nonfiction narrative about going down to the Gulf in 2010 to investigate the BP oil spill. In the section I'd read the night before, Dave had described a causeway in Alabama that "served as a dividing line between freshwater and salt . . . and doesn't allow for interaction between fresh and saltwater habitats." Dave then summed up the problem as being one of "deep connectivity issues."

And looking ahead on the Tamiani Trail, that's exactly what I was seeing: fresh water to the right, salt water to the left. We were on the manmade foundation separating these ecosystems, and so were the construction crews building a raised bridge parallel to the old road.

"This work has been going on for ten years," Jim told us, then added that he was glad, because the road we were on was basically keeping two natural halves of the environment apart.

If Florida was consciously trying to correct the mistake it had made almost a century ago, it didn't occur to me to ask. The main thing was that the state was shooting to revise the "deep connectivity issues" it had created in the name of mass transit. Not only that, there were booms floating around all areas being dug up, to hold back the sediment—which would never happen in Arkansas.

There's currently construction along Lake Conway, where the interstate is being widened for commuters going back and forth to Little Rock. When the city planners met last fall to talk to the Lake Conway Advisory Board about adding two more lanes, someone brought up the fact that there was no mention in their plans regarding run-off. However, the plans did state that the lake would not be affected.

As my university's representative, this was something I had to challenge. When construction happens, so does run-off and turbidity, which are forms of pollution. So, like usual, I spoke up. And, like usual, my input was regarded as the annoying whining of the liberal professor in a conservative state. When I said that our desire to accommodate all those cars out there creates waste that gets flushed directly into the lake, the main city planner shot me down by replying, "Cars don't poop!"

His buddies, of course, laughed at that, while I sat there actually seeing red. Because the fact is cars *do* poop. That is, the infrastructure they require does in fact soil the water, the fish, the plant life, and the complex web of microorganics that are jeopardized when bulldozers rumble in.

Here, though, in Florida, it seemed like somebody was actually trying to protect a natural treasure. I might be wrong about this, I don't know, but that's what I thought as we kept on cutting across the land. Or marsh. Or canalways growing thick with melaleuca trees and Australian pines from down under and the South Pacific. They were mixed in with Brazilian peppers, hyacinths, cypresses, everything. In that tangled cantata of roots and brush and brackish streams, I envisioned a colorful cocktail of homegrown and exotic species swimming all around us: millions of South American oscars, Asian snakeheads, Israeli tilapia, cichlids from Mexico, and walking catfish from Thailand running with the gators and gars. Because that's what Florida has become. Like its diverse human demographics, this wild and snaky sunshine state is filled with pythons from Burma, panthers from Texas, iguanas from Central America, and birds from all over the world—like South American parakeets, European and African swamp hens, and accidental occidental myna birds.

But in that moment, this melting pot appeared to be working—or so it seemed from the passenger seat of Jim's van. I mean, there were still plenty of native fish in the system, and native plants, and amphibians endemic to the region. So maybe those booms weren't holding stuff back. Maybe they were holding stuff in—like *what is*, before it becomes another thing. A less familiar thing—which we have no concept of how to handle.

I'm sure there are experts who won't agree with this assessment, who might look at me as naïve for not decrying the danger of invasive species. I guess I feel that it's best to avoid an alarmist stance. Snakeheads are loose in the United States and they aren't devouring everything as we initially feared they would. Similarly, bighead and silver carp have made it to the tributaries of the Great Lakes, and the Mississippi River system isn't totally devastated. Zebra mussels and Eurasian milfoil have also spread far and wide, and it's been a fight, but we're still kicking. So I think there's a balance we can strike in informed and constructive ways.

For example, as an August 12, 2011, article in the *New York Times* noted, commercial markets are being developed in Turkey and Indonesia for bighead and silver carp from Illinois. And out in Utah and Wyoming, they're making lemonade with illegally introduced burbot through fishing contests which help remove them from Flaming Gorge Reservoir. This process is admittedly slow, but the word is getting out that burbot are delicious, which brings in tourists and attracts more anglers who remove more burbot from the Green River system.

Change is imminent, ecosystems are always evolving, and flexibility counts. The trick is to roll with it—*it* being invasive species. Because that's the way it's always been. There's no guarantee that species will stay in their natural habitats, and humans are the quintessential example of this. We're everywhere, adapting and nesting and reproducing like we own the place.

And speaking of "owning" a place, my friend Dave, the environmental writer, is always spouting off about the value of committing to a place, living in a place, and knowing a place all your life. In eco-nonfiction, *place* is a common theme. It's one that's associated with monogamy—which I'm all for. But when it comes to the idea of being faithful to a specific place, I don't think this is practical anymore. Not in this high-speed world of jets and trains and industrial tourism and going where the jobs are. We've evolved into a species with the most wide-ranging range on earth. These days, it's not natural to stay in one place. Like the peacock bass from the Amazon, and the Florida gar in Canada, our nature is to go where petroleum takes us. And as we continue to evolve, I'm sure we'll become even more multi-abodal, humans and creatures and plant life alike.

I see myself as a Midwesterner living in the South, but with strong ties to the Rocky Mountains, the Pacific Northwest, and sporadic places in between. At a cellular level, my genes feel just as comfortable on the lakes of Minnesota as they do in the swamps of Louisiana. And with all that Spanish moss out there in the Everglades and the ibises above gliding on invisible gyres, this place was beginning to seem like a strange strong place I could adapt to, a place in which my DNA knows it could live and fish and run around for years—which is where I think the world is at.

Such are the detours of the eco-mind, whose nature it is to explore. We reach for space, we descend the depths, and essentially, humans are just as invasive as any species. Our habitats are now all over the place, but that doesn't mean all species are in competition. Species have always been clashing, merging, hybridizing, and dying out every day. It's that whole damn circle-of-life cliché. Everything is born to die.

* * *

But we are also born to live. I know this in my core when Jim pulls off on a side road in the Big Cypress National Preserve, and suddenly I'm seeing alligators everywhere. They're just lazing in the sun, exciting my inner-kid. It's like I want to clap my hands, or jump up and down every time I see one. In a sense, it's like going back in time to the days of gigantic Jurassic lizards.

We pull over a few times, get out at bridges, take a gander in the water. The streams are always clear and pure, and the Florida gar are always there, swimming in groups of six to ten. You can see their details down to the scuffs on the ends of their beaks, the scratches on their backs, their stripes, their spots, eyeballs, nostrils, everything. Some are dark, some are light, some are gray, and some are upwards of a yard long. Like us, they vary. And like us, they prefer certain times to eat.

We cast a few flies, but like Jim says, if they don't bite within five casts, we're wasting our time if we stick around. So we keep on trucking, and getting out, and gar-spotting. And sometimes there are alligators: ten, eight, six feet away. They don't bother us, and we don't bother them.

But I do bother one when I'm bringing in a fly. It's just hovering there, a three-footer, looking out at the lily pads. And here comes my fly. It's three feet away, two feet away, one foot away. I know it's not ethical, but I pull my line right over its head, then let the fly sink. The gator's hide is so tough I know it won't hurt, and then I start reeling in. When the fly snags the alligator's back, I crank back and set the hook.

Holy Crap! The water erupts, the gator takes off, and suddenly I'm hauling back on a thirty-pound reptile. Tom comes running, and Jim is concerned that my rod will break. "Lower The Tip!" he yells to me.

I do, then strip in the line a foot at a time as the alligator dives and the battle rages. It's trying to swim away from me, body swishing snakily. I even break out my camera and take a few shots while horsing it in, before getting back to the business of what to do next. Because it's beneath me now, and all I have to do is reach down and grab it by the tail.

I'm confident I can handle this. If it snaps at me, I'll drop it like a hot potato—or an alligator trying to chomp off my hand. If I get it on land, I figure I'll grab it behind the head. But then, of course, the line snaps, and everyone is better off.

"Don't put those pictures on the Internet," Jim says, knowing he could get in trouble.

Later on, I'll look up the regulations and find out that "harassing" gators is not allowed.

If what I did can be categorized as harassment, though, I don't know. To me, in that instant, it felt like I was totally alive and part of my environment. Maybe if I'd called it some abusive names and tried to demean it, then that would've been a form of harassment. But whatever the case, I can definitely say that there are few thrills in life like fighting an American alligator on a flimsy little fly rod.

Photo 27. Gator On!

Photo by Mark Spitzer.

There were three other notable things that happened in Florida.

First, Jim took us to "The Skunk Ape Research Center," which was actually a gift shop run by this guy who's been mucking around in the bogs for years, trying to prove such primates exist. It was a fun stop, and that guy talked our ears off, telling us how he'd taken some famous video footage, and how the Discovery Channel had him under contract.

I ended up buying a T-shirt and a beer coozie, and he threw in a bumper sticker for free.

Though I wasn't a card-carrying member of the Bigfoot Society of America, I'd done extensive research on the iconography of the inter-section of our wild and "civilized" selves, which has been a theme in art and literature since the *Gilgamesh Epic,* four thousand years ago. Back in college, I had formally studied the concept of the Medieval and Renaissance "wildman" in depth, resulting in a four-hundred page senior thesis that looked at such missing links through the lens of art history, world literature, and psychology. More recently, though, I'd been highly involved in a cryptozoological study of mythical creatures in Arkansas, based on folklore, science, interviews, etc. A big part of that Arkansas project focused on sightings of what I called the "Arkasquatch." This led to a conversation on the way back in which Jim pointed out the window and said that if skunk apes were really real, and if they were really out there, then they'd sure have a lot of room to slog around, since there was nothing from here to the coast except sixty miles of undeveloped salt marshes.

The second notable thing that happened in Florida had to do with a cool little pickerel Tom and our wives and I saw the next day in a city park canal. It was about ten inches long and didn't seem to care that we were standing right over it. Then we saw it lunge. It nabbed a smaller fish, then snarfed it down just like that—which is something you hardly ever witness in the wild.

The third notable thing that happened was on the Ft. Lauderdale Beach later that day. Robin and I were having a beer and watching the sunset, and there were these pint-sized sandpipers skittering all around, pecking at the beach when the waves receded. When the waves came in, they always cut it close, escaping the tide by just a few inches. Sometimes, however, they'd get caught in the backwash, but they always managed to work themselves free and scuttle back up the beach. There was one sandpiper, though, that wasn't so fortunate. It skittered too close to an incoming wave, which crashed down on it and swept it out to sea right before our very eyes.

We could barely believe it, that we'd been there to see that scene. I mean, does this happen all the time, or was that particular piper just not destined to carry on its genes?

All this happened at a time that I was trying to make sense of my Florida gar experience. In a sense, I was looking for a way to frame my "investigation" as I tend to do with the adventures I turn into chapters for this book.

In the case of the pickerel that took down that minnow and that not-so-bright sandpiper, the symbolism was obvious. Too obvious, in fact. Making a statement about the brevity of life and how it can get snuffed out so quickly has been done a million times. To force Florida gar into this scenario would be old hat, and even insincere. That's not what this trip was about.

The skunk ape idea held more promise, but I still wasn't sure how to use it. There was all that swampage out there, full of legion native and non-native species living together just like us. But again, the idea of manipulating that into some sort of moral to the story just didn't feel right.

Still, I had a thought in the back of my mind. It was absurd, but I figured it might be appropriate and add finality. Plus, I was curious, so just had to try it out.

* * *

Back in Arkansas, I did a bit more research on gar as a substitute for lobster and found out that the article I'd read had missed the point. There were a number of recipes on the Internet for "redneck lobster" or "poor man's lobster" or "lake lobster," but they never intended for gar to be a substitute for the real thing. If anything, these recipes were playing with the idea of gar having a lobsterific quality.

So when a three-and-a-half-foot longnose appeared beneath my boat one day, oblivious to the fact that I was there, I wanted it so badly that I literally hurled myself over the rail, plunged both arms into the river, and grabbed it with both hands. I got it in an iron grip right beneath its pectoral fins and pulled it into the boat. It weighed 6.5 pounds and I brought it home, cut it up, then set a pot of water to boil with half a pack of crab boil in it. I threw in some potatoes and carrots and some garlic cloves, turned it down to a low boil, and let that cook for twenty minutes before throwing in the gar chunks and onion slices. I even threw in a lobster for comparison, let it all simmer for five more minutes, then turned off the heat, covered the pot, and let it sit for five minutes.

This concoction was pretty similar to a faux crab recipe I've been making for years with freshwater drum, which I love but Robin hates because of its rubbery texture. The only difference between these dishes is that after straining, you serve the drum with cocktail sauce, whereas the "gar lobster" gets dipped in melted butter. So that's what we did, and the lobster was exquisite.

As for the gar, it was nothing like the lobster at all, but it was damn good! The consistency was a bit chewy, but Robin said it was better than gou.

Meanwhile, the world wasn't any closer to resolving any of its deep connectivity issues, but at least I'd discovered a new way to cook gar —which is about as much as anyone can hope for and achieve in this crazy flux of flora and fauna we tend to see as a chaotic network of overlapping habitats. But guess what? It's all one vast ecosystem, and if something doesn't seem natural, it's only because we refuse to accept that there are no borders in the wild.

Photo 28. San Carlos, Nicaragua.

Photo by Mark Spitzer.

Chapter 8

First-World Problems in Third-World Countries

Trolling for Tropical Gar

So I lit off for Nicaragua to investigate the most mysterious gar: *Atractosteus tropicus*, alias the tropical gar. Compared to other gars, there's just not much literal knowledge about this fish. For instance, the Gar Anglers Sporting Society (GASS) website notes, "The Truth is still murky . . . . Maximum size? Unknown. Clearly they get as big as their close cousins, our alligator gar. Alligator gar of over 300 pounds have been documented." But according to our limited info, the heaviest known tropical gar on record is 6.4 pounds, and they rarely exceed 1.25 meters in the wild—so clearly, they don't get as big as gator gar.

Anyway, since my wife wasn't about to let me go off alone on such an exotic adventure, she was along for the ride—to the Rio San Juan, which marks Nicaragua's border with Costa Rica, as well as the southernmost known population of tropical gar in the country. The GASS website recommended the guide service San Carlos Sport Fishing, and since they were accredited with the International Game and Fish Association (IGFA), we had purchased an all-inclusive fishing trip based out of the

Jungle River Lodge on Lake Nicaragua. This five-night six-day package included air-transportation from Managua, fishing licenses, guides, bait, boat, tackle, accommodations, three meals a day, and beer.

We were met at the airport by Philippe Tisseaux, who ran the show. He was a sixty-year-old French citizen and veteran angler flanked by his teenage son and daughter, and we all piled into his SUV for a four-hour trip through mountains and deserts and cattle-donkey-horse-covered roads where dirt-poor peasants traversed the shoulders in the blazing sun, transfixed by cell phones as swerving trucks and buses rushed by.

This wasn't the small plane ride we had expected, but it allowed us to see the countryside, and it allowed me to question Philippe about catching gar. Whereas the GASS site noted Philippe's outfit averages "five gar per outing" with the "exciting 'by-catch'" on these excursions being "massive tarpon or snook," Philippe told me that when he first opened his business they caught five gar per tarpon—but now, a decade later, it's five tarpon per gar. Whatever the case, the take-away message was that gar were caught by accident trolling for trophy tarpon in the 100- to 200-pound range. According to Philippe's website, most of these gar weighed ten to forty pounds, and according to Philippe, their numbers were in severe decline.

So we traveled by car, then traveled by boat, and eventually got to the jungle lodge, located on La Punta del Diablo, a spit bordered by the Rio Frio on one side, and a vast freshwater horizon on the other misting into distant volcanoes. Robin, however, wasn't very impressed by the "luxury room" we had been promised (no towels or bottled water as stated on the site), and I was disappointed that there was no beer from Philippe's alleged "private bar," which was nothing more than a kitchen fridge that was inaccessible to his guests.

A few hours later, the beer arrived, and we sat down for an unidentified barbequed mammal that Philippe had purchased from a kid on the side of the road. This jumbo drumstick had traveled with us at room temperature all afternoon and was supposedly deer, but it could've been a big goat or it could've been a small horse. Robin was wary, and thought it was greasy, but we gobbled it down and didn't complain—to our host.

After all, as privileged tourists crossing the land, we'd seen the crippling poverty, which always comes stocked with malnutrition and disease. In thousands of shacks surrounding us, shredded plastic flapping in the wind, emaciated children were going to bed with fevers and lesions and no education whatsoever. Our inconveniences, therefore, were merely "First-World problems." And ultimately, we weren't there for comfort and amenities; we were there for tropical gar.

* * *

Robin and I took off at 7:00 a.m. with our guides, Mino and Andre. There were howler monkeys along the way, swinging from the branches of fruit trees, and iguanas and ospreys and hot-pink roseate spoonbills. In fact, there were thousands of exotic and not so exotic birds, as well as bright and brilliant complex flowers bursting from the ferny, viney, jungle brush. Heading up the Rio San Juan, we saw water plants, coconuts, what looked like kudzu consuming trees, and tarpaper shacks with satellite dishes pointed toward the sky. Nicaragua on the left, Costa Rica on the right.

The idea was to start out with live bait, fishing off the bottom, so we stopped to gather some baitfish. Mino used a cast net and brought in a bunch of primitive-looking armored catfish with suction-cup mouths and ornate fins. They called this fish *pleco,* then threw them on the shore like all bullheads everywhere that are hated by humans. I looked this literal sucker up later and found out it's considered a territorial pest.

Photo 29. Pleco.

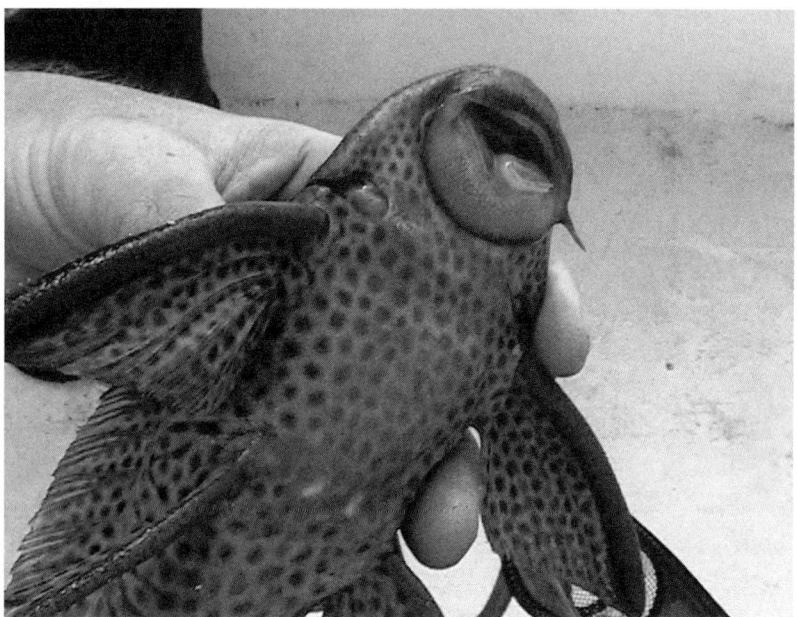

Photo by Mark Spitzer.

We also caught a lot of cichlids, which resemble sunfish, and big-ass mosquito fish, and shad-looking fish of all kinds. Each cast brought in a strange, new, bright-colored specie.

With our bucket full, we headed on to a highly garful-looking spot with a slow-moving muddy current and tall grass along the shore, kingfishers diving everywhere. Then casting out, we settled in, waiting for gar, Arkansas-style.

That's when I noticed that our guides didn't know squat about fishing for gar, even though the GASS website claims that Philippe's outfit targets this specific fish. If our guides knew anything about garfishing, they would've left their bales open, to allow the gar to run with the bait. Also, from a conversation the night before, I'd learned from Philippe that neither he nor his guides knew the difference between tropical and alligator gar—which they basically considered the same fish: *el gaspar*.

Nevertheless, I saw a few rise, but I also saw a tarpon rise. This was a fish I knew nothing about, but when its head came up, followed by a sharky fin, and then another down by the tail, I glimpsed its girth. That fish was six feet long and weighed at least a hundred pounds.

And a couple hours later, after getting skunked on the bottom, I latched on to one. We'd switched to trolling and we were having a lunch of ham sandwiches and potato salad. I was sitting on a lawnchair in the bow when a heavy-duty rod bowed and Andre hit the gas, setting the hook. That's when it broke the surface, bursting twelve feet into the air. Then it hung there, six feet above the water and six feet from head to tail, pointing straight into the zenith, shuddering silver in the sky. By the time it splashed back down, I was also hooked.

That fish was a game changer. Suddenly, there was another strong reason to be in this place—a reason that instantly shifted the focus of my mission here.

Mino handed me the rod and we all started reeling in fast—the others to get their lines out of the way, and me to bring that badboy in. Mino, however, motioned for me to slow down, and said something that Robin translated as "no mucho pressure." Between her lapsed Spanish and my crappy French, our communication was limited to a pretty sparse vocabulary. Detailed instructions were not possible. Gestures were employed frequently, including a lot of shoulder shrugs.

Anyhow, I kept on reeling, and that fish kept taking out line. I'd bring in five yards and it would take off with twenty, spinning that drag in a frantic frenzy. It was a lightweight Shimano baitcasting reel, a company I was familiar with due to the parts on Huffy bikes I always had to replace as a kid. Those bicycles were also pieces of crap.

Five minutes into the fight, I was exhausted, but I knew I couldn't give up. I mean, if Hemingway's son in *Islands in the Stream* could fight an epic monsterfish battle all night long, what sort of wuss would I be to ask for assistance from our guides? Little did I know, I still had forty furious minutes left.

In the second five minutes, I felt its weight. That fish was stronger than any six-foot gar I'd ever played, pulling with about two hundred pounds of sheer force. Hence, I had defaulted to the pull-back-then-reel-in technique I'd learned from hauling in sturgeon in Oregon, which was necessary because the reel was so lame. It just couldn't handle the size of the fish, sometimes giving up thirty to forty yards at a time.

In the third five minutes, Mino gestured for me to lower the rod, so that if something snapped I wouldn't go flying back and smack someone upside the head. I understood this, but what I couldn't understand was why he kept suggesting that I let the fish take as much line as it dang well pleased when the idea is to reel it in. I decided to try to slow its run, but when I pressed down on the woven line, that tarpon just burned the flesh off my thumb. So I wet my hat and tried to use that to slow the spool. That also didn't work.

Twenty minutes into the fight, my back was sore and the butt of the rod was eating into my pelvis as the tarpon tugged like an enraged bull. I'd bring in ten, it'd take out thirty. Relentlessly! Consistently! Sometimes jumping, allowing us to fully gawk at it—maybe even seven feet long! Landing, splashing, yanking, thrashing!

I started making progress, even though the reel was falling apart. It had started out with a missing knob, and I had tightened the star drag to the max, but that fish kept taking out line. It was the strongest fish I ever fought, and the reel was now totally fried. I watched as it took off downstream with at least sixty yards I'd worked twenty minutes to reel in.

Mino came over and adjusted the handle, which had come loose. Still, when I turned the handle, line kept going out. So I started turning the drag instead, which, in turn, turned the spool, which was maddening.

Because I could only turn it a notch at a time, which would only bring in a few inches at a time.

This method, however, combined with a technique I made up on the spot—of walking backwards on the boat, then walking forward while reeling in—worked well when we chased the fish downstream.

By the twenty-five-minute mark, the neon green leader was visible and we had the tarpon next to the hull. It had scales the size of beer coasters and I figured it weighed at least 120 pounds. The oversized Rapala lure was snagged on its dorsal fin and pointing toward its tail, so the tarpon was swimming away from the boat.

Mino got down on his belly in the bow and grabbed the leader, and the tarpon shot off. He held on and horsed it back, but then the fish switched directions, blasting to the right side. Mino rolled with it and I found myself standing above him, the fish pulling straight down. Mino was clinging to it with one hand and looking up at me like "HOLD ON, MAN! HOLD ON!"

But we couldn't hold on, neither of us. That fish torpedoed off and it took another ten minutes to bring it back. Mino grabbed the leader again, pulled up, and suddenly... nothing but the lure was dangling there. We stood frozen with our mouths agape, and I could see in Mino's eyes that he'd wanted that fish as much as me.

But what can you do? That tarpon got off—so now I had another fish to catch.

Photo 30. Guardian of the Gar.

Photo by Mark Spitzer.

The second day began with an anteater hissing on a pile of gar, its arms outstretched like a menacing angel. When the fisherman pulled up in his dugout canoe, the staff, knowing I was there for gar, had called me down to the dock. So there I was, trying to take pictures of gar while a monkey-possum-looking-creature threatened to claw my face off.

We'd watched the fishermen the night before, spreading their nets at sunset in the shallows of Lake Nicaragua, which was a hundred miles long and half as wide. Philippe had explained that the gar they caught overnight would be sold at the local market in San Carlos, whereas larger gar, caught deeper in the lake, would be dried and shipped to Guatemala, which was helping to deplete the local population. Basically, there were no regulations. When Philippe first came down here in the nineties, he claimed there'd been huge gar all over the place, thousands and thousands and thousands of them, jamming the river. Now, however, the gar were smaller, and a lot less abundant—a story we're familiar with up in the United States.

According to the December 2008 issue of *Journal of Fish Biology*, tropical gar have been disappearing from Mexico and Central America due to "habitat degradation and destruction." But looking beneath that anteater, I could see another major cause. The paler ones were alligator gar, and the darker ones were tropical, and none of them were more than a foot and a half long. There were about seven gar total in the boat, all of them skinny and snaky and too young to reproduce. With pressure like this, it was pretty clear that overharvesting was a determining factor in the decimation of this niche.

This observation was reconfirmed a few hours later when we pulled up on the shore of *La Isla del Gaspar*. It was out in the lake, a small island that was part of the Solentiname archipelago system, and it was covered with gar heads, because that's where fishermen process large gar.

There was another boat already parked there and a kid with a Spiderman shirt was curing gar. There was a split-open gator gar drying on a plank, its flesh coated in salt from a sack on the sand beside it, and there were chunks of gar in a bowl. And all along the shore, there were alligator gar heads and scaly hides curling in the sun.

Photo 31. La Isla del Gaspar.

Photo 32. La Isla del Gaspar.

Photo 33. La Isla del Gaspar.

Photos by Mark Spitzer.

These were the big ones, but they weren't that big. From the size of their heads and the lengths of their skins, I estimated that most of these gar were between three and four feet long, meaning they were also barely old enough to spawn. They'd been caught in gillnets and conked on the head (most of them had fractured skulls), and would soon be shipped to Guatemala to supply a demand that was leaving the largest freshwater lake in Central America with a diminishing fishery.

We spent the rest of that morning casting in the cacophony of one of the most roostful islands of waterbirds Robin and I had ever seen: blue herons, brown herons, green herons, techni-colored checkered herons, plus cormorants, snowy egrets, ibises, et al.

Then we went to a marshy bayou and went motoring up and down, looking for gar in another natural habitat. We saw a few rise here and there, but never got a good glimpse. We definitely saw more garfishermen than gar, spreading their homemade pop-bottle-float nets.

* * *

After lunch and a siesta, we were back on the San Juan, trolling for tarpon. I had the fever now, and had to get one—and since this was how gar get caught, it made sense to embrace this method.

The only problem was the triple-hot sun, which had roasted us the day before, only twelve degrees above the Equator. Though the temperature was in the eighties, it felt well over 110 between the hours of ten and two. We'd worn short sleeves the day before, but had donned long sleeves today, which seemed ridiculous, but the sweat was worth it. Even our guides kept shifting around beneath the six- by twelve-foot canopy, hiding out from the frying rays.

We hooked one tarpon and I fought it for twenty minutes on a spinning reel, but again, it was taking out too much line. So I tried to stop it by slowing the spin of the spool cover. The line got too tight, and then it snapped, making off with the lure.

But at least we caught a fifteen-pound snook right before the sun set. It was a beautiful, fat, bassy-looking fish with a prominent slooping underjaw, which the chef at the lodge prepared in a glazy, garlicky sauce.

Photo 34. Snook.

Photo by Robin Becker.

We ate it that night with Philippe, when he finally informed us about the proper method for bringing in tarpon. Supposedly, you're just supposed to let them run. And run and run and run and run and run and run and run and run—for at least an hour. Because the line was only thirty- or forty-pound test, because that's how they do it. If you try to slow a tarpon, the line will break.

"You need to respect the guides," he told us. "You must listen to them."

To which Robin responded, "We did, but we couldn't understand them, because there's a language barrier."

"No no no," Philippe replied with a finger wag. "Language is not a problem."

He then explained how Mino and Andre could communicate anything that needed to be communicated, so if they weren't doing that, then they weren't doing their jobs.

Anyway, we gave up on that conversation, and later that night, we heard Philippe yelling at our guides. He'd done this the day before after I told him how we'd lost that first tarpon thanks to that crappy reel. But here's the thing:

If I'd known that line was so lightweight, I never would've applied so much pressure. Sure, our guides had warned me, but since language *was a problem*, they hadn't been able to tell me why. And Philippe could've told me *why* two days ago. And he could've told me *how*. And because of this miscommunication—or *dis*communication, rather, since it was essentially *dys*functional—we'd lost two huge fish, and still no gar.

Meaning the pressure was on, tensions were high, and Robin and I, with our First-World problems, were becoming increasingly annoyed at trivial things. Like the surly kitchen lady who sneered when we asked for a beer. Or Philippe's bratty teenage son complaining about the clientele. And how about a blanket for our bed—since we're paying thousands of dollars to be held hostage, in a sense, in this Colonial, down-the-Congo, Gilligan-hut tourist-gulag? Plus, maybe we'd like to go into town and get a bottle of rum, since Philippe's "private bar" seems to be more private than advertised, et cetera, et cetera.

But at least the evening was cooling down and the mosquitoes were pretty much non-existent. Nevertheless, we went to bed listing our grievances to each other, already planning a scathing review for *Trip Advisor*, frogs burbling in the night.

* * *

The third day was hardcore, trolling all day in the broasting sun. We had five lines in the water: four in holders, one in my hands. On those rods, there were three big lures for tarpon, and two smaller ones to increase our chance of getting gar—which, by noon, I'd essentially given up on. We were definitely getting more tarpon hits, so that's what I was shooting for. If we caught a gar by accident, so much the better.

Then it was time for lunch, but guess what? There was no lunch on board, just some crackers and yogurt our guides brought along and shared with us. So what was the meaning of this? Were we getting punished for using "too much gas"—which Philippe complained about the day before when we'd motored around all over the lake and marsh, then went up the San Juan? Or were our guides being punished for losing lures, and not instructing us properly? We figured the latter—that Philippe had ordered the guides to supply the grub, and they, being poor, just didn't have the cash on hand.

So we kept on trolling through the broiling afternoon. Robin read a whole book and even reviewed it while I shifted positions in search of shade, always holding tight to my rod.

By four o'clock, we were all burnt out, when Mino said, "Uno más," and indicated one last circuit of the mile-long stretch we'd been cycling through for half a day. "Okay," I nodded. And then it hit.

Bringing it in, I knew it was a gar: the easy way it came through the water, the lack of yank compared to tarpon. Then we saw its tubular shape: a bit larger than two feet long. All I did was lift it in the boat, and it was ours!

POW! Cheering ensued, then jumping up and down, and high fives all around. It was like divine intervention. We broke out the cameras and started taking shots. And with no seething anteater guarding this gar, I was able to take a closer look.

It was a tropical gar, just what I had come to get, a healthy fatty with dark gray coloring on top and stripe-like markings down by the tail rather than spots. The tail skin was thick and almost opaque, colored as darkly as its back, along with its rear fins. It also had a flat, broad head similar to an alligator gar, but a bit more slim, like spotted are. And when we took the eight-inch floating Rapala from its mouth, I saw two rows of teeth on the upper jaw, just like gator gar.

Photo 35. Mark and Tropical Gar.

Photo 36. Mark, Mino and Tropical Gar.

Photos by Robin Becker.

The biggest difference, though, between this gar and the other five species I'd caught so far, was that this one talked. Yep, I was surprised to find that tropical gar have a voice like a frightened catfish chortling out a gurgling sound that comes from exhaling air. It sounded like a groan, or a grunt, or an old dog sighing with a snorey sound.

I decided we'd eat it. I mean, why not? Philippe had told us how his chef cooks gar with "*real* mayonnaise." So when in Rome, I figured, we should eat a gar.

But as we motored back for another "uno más" circuit, I considered the state of this fishery, which I'd already made my judgments about.

The gar here were running low, and though taking one out of the mix wouldn't really have much effect, it was the principle of not indulging when I could that kept bugging me and eventually got to me. I just couldn't sacrifice this gasping gaspar.

"Gar va," I told our guides, hoping this meant "gar go." Then I added, "gar libre!" figuring this meant "gar free!" Andre laughed, stopped the boat, and I lowered my catch over the rail, where I held it for a few minutes. When it was strong enough to swim off on its own, I called out "Adios, Señor Gar!" and we watched it swish into the murk.

Photo 37. Gar Libre.

Photo by Robin Becker.

Now that my mission had been accomplished, this took pressure off everyone. When we got back to the jungle lodge, Philippe was so enthusiastic that he called for a bottle of Nicaraguan wine. He poured glasses for us and our guides, raised one himself, and made a toast to *El Gaspar*—so it was hard to remain mad at him for all our piddly First-World concerns. Like why hasn't he arranged for our flight back to Managua, like he said he'd do three days ago? And why hasn't he called our hotel there, to let them know when we'd be at the airport, like he also said he'd do? Nope, those issues just ceased to exist as we drank our sweet wine in the cool breeze of the night.

Our charming host then made another toast, this one to the next day's catch. "Mañana," he told us, "we will finish this bottle when you catch a tarpon. And we will also eat a gar for dinner. I will buy one at the local market, and prepare it with *real* mayonnaise."

Photo 38. Basilisk Lizard.

Photo by Mark Spitzer.

But we didn't catch a tarpon—they didn't even bite. It was just "too muy hot," according to our guides. We did catch three snook, though, which we threw back as we trolled all day in the blistering sun. Robin read another book, and then it was time for lunch.

Again, we'd left the lodge without any grub, but when we stopped in town to get our manifest stamped, I'd given our guides some dough for *tortas*. They bought ham, bread, pastries and Coke, which we ate in the shade of an orange-flowered tree, a bright green basilisk lizard lounging in the mangrovey roots.

When I got out of the boat to take a leak, I stepped foot in Costa Rica. Needless to say, there were no custom agents around or forms to declare why I was there. Then, just as quickly, I was back in the boat, casting out my Rapala.

All afternoon we motored around, not catching anything more than sunburn. But it didn't feel like this was the end. It felt more like a beginning. Now that I knew tarpon existed, I knew I'd try again—somewhere, somehow.

When we got back to the lodge, however, Philippe told me that I'd already caught a 120-pounder.

"What do you mean?" I asked, and he replied that both Mino and Andre agreed that the tarpon I caught the first day weighed 120.

"But I didn't catch one," I replied.

Philippe then explained that according to IGFA rules, if someone on the boat touches a leader with a tarpon on it, then it's officially caught in this neck of the woods. The logic, he went on, was that since tarpon are protected, you're not supposed to bring them into the boat, because taking them out of their low-gravity habitat can damage their organs. That's why they don't use gaffs, and that's why all the trophy shots around the lodge were of people hanging over the bow with their hands down a tarpon's throat.

But I wasn't buying it. I never touched that fish and we never got a photo of it under anyone's control.

"Whatever," I said, then sat down for dinner, which wasn't gar with *real* mayonnaise. Which was fine with me, since I wasn't sure I wanted to play a part—no matter how miniscule—in contributing to the weakening of this already compromised ecosystem. Still, it was frustrating to be promised something, then have that forgotten the very next day.

* * *

In the morning, we waited half a day for our change and receipt. Since Philippe never booked our flight, we also had to wait for a guy to drive us four hours back to Managua. And on the way, he fell asleep—twice. The first time he crossed the center line and the second time he drifted onto the shoulder, almost mowing down a girl with a fruit juice bottle, running toward us to make a sale.

The flight the next day was just as hectic, with some pushy ladies shoving their way to the front of the plane when we landed in Houston. I jumped up and got in line according to standard disembarking protocol. This led to a heated verbal skirmish in which both parties, experiencing typical traveling tensions, allowed the worst in themselves to boil over.

Then, at the Little Rock Airport, my luggage was missing. After an hour and a half of being inconvenienced, we arranged for it to be delivered. My car key, however, was in that bag, so out in the rain, I had to crawl beneath my Jeep and lie in a puddle to retrieve the hidden key. I was so soaked that I threw my shirt in the backseat and startled the parking attendant, who thought I was some sort of creepy late-night nudist pulling up to get his kicks.

To complicate things even more, our parking ticket wouldn't swipe, so she entered the info by hand and ended up charging us $35 extra, which I realized after she'd made change. It took her twenty-something irritating minutes to fix this mistake, while Robin and I sat there with the engine idling, definitely back in the USA.

But we finally made it back to our First-World house, where collapsing on the Ikea couch and turning on the HD TV, I liked the idea that down in Central America the Tropical Gar groans for Man. From what we'd seen, though, I knew it should be the other way around. That is, due to an economic desperation in which Third-World people sweat and scratch and starve every day, Man should really be groaning for Gar—whose highly exploited populations are going down, thanks to a dearth of metaphorical anteaters sticking up for *El Gaspar*.

Photo 39. Photo from eBay.

Chapter 9

Thailand's Lake-Monster Fisheries

Investigating Gator Gar and Arapaima

My mission in Asia was three-fold: 1) to investigate the concept of "fishing parks" in Thailand, 2) to catch an alligator gar, and 3) to catch an arapaima, which is a massive, prehistoric, boneheaded fish from South America. That's why I had signed on with Siam Fishing Tours, whose website boasted gator gar and other top predators, such as mongo Mekong catfish, mega giant snakeheads, behemoth barramundi, cattle-sized carp, and the already mentioned monster-fish, arapaima. Such super-sized trophy fish had been put into man-made lakes surrounded by huts you can order food and drinks from while catching exotic species all day. These fisheries are becoming increasingly popular in Thailand, especially with Australians, Canadians, and European sport fishermen.

Robin and I were picked up in Bangkok at 6:00 a.m., then shuttled down to the southwest tendril of the country for one of the most unusual angling adventures we'd ever been on. We arrived at Jurassic Mountain in less than three hours, and before nine, we were on the lake, which was situated amongst rice paddies, dramatic straight-up-into-the-sky mountains rising through the mist around us, bejungled with vines and coconut trees. The five-meter-deep lake was only the size of two city blocks, but there were jumbo species everywhere. I saw one of the two six-foot gator gars roll, but that was nothing compared to the other creatures coming up in the lake. It was full of 200- to 300-pound arapaimas, some almost nine feet long, surfacing like submarines. They'd shove the water away with tsunami force, and then you'd see their backs rise. And continue to rise. And just when you thought it was all over, here'd come its dorsal fin, and then the tail. Some were an electric neon mauve, and others were an ominous black.

The manicured, pebbled path was surrounded by bright orange orchids and other flowers, and there were storks flying from pond to pond. We had a bamboo hut with a cooler full of ice-cold drinks and pillows to lounge out on, and the service of a "gillie," who baited our hooks with thawed-out mackerel and fishmeal pellets, then cast out for us and placed our rods in holders equipped with sonic alarms.

It was weird to have this done for us, and I was worried that not getting dirty in pursuit of fish might be a form of cheating. I was also skeptical of the idea of "fishing parks." Like was it also a form of cheating to have all those fish in a barrel right in front of us, pre-packaged in a sense, while a Third-World peasant chummed for us? But when that first fish struck and I set the hook, I threw my back into the task. Suddenly, I had my own job to do, and I had my own skill to apply, while our jolly gillie advised on the proper technique. As with tarpon down in Nicaragua, light braided line was being employed on big-game rods and reels, the idea being to wear the fish out by letting it play the drag. Hence, I was hauling back, then reeling in as I lowered—because somewhere in that body of water, there was a delicate but enormous creature attached to me by a stout barbless hook no larger than a quarter.

The owners of the lake were serious about protecting their investment. So when a carp mouth rose large enough to swallow a softball, followed by a forehead the size of a basketball, I let the drag do the work. Eventually our gillie tightened it up and I pulled the fish into the net. It took both of us to lift it from the lake and place it on the mat, which was there to insure it wouldn't get injured when we took out the hook. It flopped around as any healthy sport fish would while we tried to keep it in the foam zone, so it wouldn't scuff itself up on the path or bruise itself against the ground. Then, when it had settled down, the idea was to keep it close to the mat when posing for the money shot, so if it leapt from your grip, it wouldn't fall far and it would have a soft landing. At that point, though, I didn't know that, so I got the shot standing up.

Photo 40. Thirty-Pound Siamese Carp.

Photo by Robin Becker.

There was one other fisherman on the lake, a horticulturist from Holland on the other side. The place was still under construction, with landscaping going on in various places, and two-thirds of the comfortable modern rooms having been constructed. The office/club house had a terrace beside it with a pond for growing stockfish. Jules Fernandez, a co-owner and British expat, had been building this place for two years, and he figured he had a year to go before it was finished, at which time he expected to do more advertising for more customers.

I found this out when Robin and I went in for lunch at the bar/café, where I found it easy to connect with our friendly host. Jules had been a fishing guide in Southeast Asia for years, but had always wanted to create a fishery of his own, so had made his dream come true.

As we ate our curried squid and pad thai, I steered the conversation toward alligator gar. Jules wasn't sure where the ones in Thailand came from. They weren't indigenous; mostly, they were fish that had outgrown their owners' aquariums.

"They don't grow very big here," Jules told me, "not much longer than two meters."

He reasoned that it might be Thailand's extremely hot climate, or they might be from Central America, where they don't grow as large as they do in the United States. I couldn't say for sure, but from the pictures I'd seen of Thai gator gar, they looked a bit snakier than those I was used to, and they had fewer spots on their tails.

Photo 41. Jules with One of His Gator Gars.

Photo Courtesy of Siam Fishing Tours.

Anyway, I'd only fished half a day, I had three and a half days left, and I was anxious to get back on the lake. Where I soon caught another thirty-pound carp, which flipped out of my arms, and thumped its head against the mat. This made me cringe, but it seemed to be alright. I'd been kneeling, so it only fell less than a foot. After that, my gillie advised me to hold tight to a pectoral fin when cradling fish.

An hour after that, I hooked another Siamese carp. This one came in at mach 10 and was twice the size of the others. I fought it for fifteen minutes, but it kept on evading the net, taking out line, and charging like a kamikaze. In the end, the barbless hook popped out of its mouth, but I was not bummed at all. Missing fish is part of the deal—in fishing parks as well as the wild.

As I was learning, fishing parks take the factor of locating fish out of the equation, but not entirely, and this doesn't lessen the experience at all. First of all, you still have to attract them, and once that's done, there are specialized techniques to apply. Also, the fact that you have to be extremely careful with these fish increases the stakes.

One can definitely make the argument that fishing in fishing parks is even more of a sport than fishing in the wild, since more of a sense of sportsmanship is involved. One thing that makes a sport a sport is a respect for one's opponent. In this case, the opponent is the fish, which you want to treat in such a way that others can enjoy it after you. This doesn't always happen in the wild, where the goal is to horse it in and get it at any cost. Then maybe you release it, or maybe you don't. A fish from the wild may never get caught again in its life, and arapaima, of course, are endangered species. According to Jules, most arapaimas caught in the Amazon die as a result of being captured. I know that when I'm fishing in the wild, I don't think much about injuring fish.

The afternoon started getting hot, so Robin wandered off, and I found myself lapsing off. I had a shady spot, and so did my gillie, just twenty yards away. At this point in the day, from noon to four o'clock, the fishing slowed, so I was told, and it was part of the culture to nap. If an alarm screeched, you could jump up and deal with the fish.

So that's what I did, and I didn't catch much. After four, though, the wind picked up and so did the fish. I caught a fifteen-pound redtail catfish that belched when I landed it, and a two-pound tambaqui, which looks like a piranha and can supposedly grow to seventy pounds.

I quit at dusk and came in. Peter from Holland had done about the same, and Jules was apologetic, claiming it was a slow day due to the muggy weather. But as we drank some beers, then shared some plates of Thai cuisine while the arapaima rose and dove (sometimes leaping in the dark, then smacking the lake with resounding mass), neither Peter or I could complain about meeting species we had never caught before.

As Jules had noted earlier, there was ten to twelve tons of bio-mass in that lake and over 1500 fish, not including the thousands more they fed on. The next day, however, I had other plans. In order to catch my target fish, Jules figured it would increase my odds to visit a nearby fishing park packed with alligator gar.

*   *   *

After breakfast, Jules drove us to Greenfield Valley, where it was reported that the water was down and the fish were more concentrated. He was lamenting the death of one of his carp, which Peter had found floating on its side early that morning. That fish was worth $400 and it was deader than a doorknob and no one knew why.

It took an hour to get to the lake, which was aboil with five-foot "alligators." That's what Jules and his buddy David Wilson, co-owner of this fishery, called gator gar. Excited by the cloud cover and light rain, the gar were in the deep end flashing belly, slapping tail, coming up every six feet. But there were other fish in there as well. Like incandescent arapaima (the F2 model, bred for the aquarium trade, then farmed into monstritude), the prized Chao Phraya catfish (160 pounds!), assorted sixty-pound carp, etc.

The gear was also similar (surf-size reels with forty-pound monofila-ment on heavy-duty predator rods), and just like at Jurassic Mountain, there was a four-ounce weight above the leader, meant to protect the arapaima. When they got a taste of metal in their mouth, or it bumped them on the nose, this let them know they better not swallow any more —which kept them from gulping down tackle. Being a lot more sensitive than gar, arapaimas could be killed by swallowing hooks. At an average of $4000 each, it made sense to protect the most valuable livestock in the lake.

Anyway, the moment my hook hit the water, something big took off with it, right when my gillie cast out my other pole. We both set our hooks, and then he insisted I take his pole. He didn't know I had a fish, but I took it from him and handed mine to Robin, leaving the spool open, so I could deal with that fish later.

The fish on the gillie's line turned out to be a lot smaller than the big one I put on hold, which I could see shooting off beneath the surface, its buffalo back shoving the water in pure bully fashion and making a wake. I hauled a seventeen-pound redtail in, then shot back to the other pole, which was now caught on a stick out in the lake. By the time my gillie paddled out there in a small boat, it had broken the line and got away.

Photo 42. One of the Many Ten- to Twenty-Five-Pound Amazon Redtails Caught in Thailand.

Photo by Robin Becker.

Jules took off and I got back to fishing. It didn't take long to catch a seven-pound pacu on a chicken gizzard. It looked like a huge, flat, silver piranha. These invasive vegetarians had decimated the 1500-mile-long ecosystem of New Guinea. Basically, those things have molars and jaws designed for cracking the shells of seeds and nuts, but because the lineup of their food chain had changed, they'd metamorphosed into vicious ball-ripping carnivores—as depicted in the "Nut Cutter" episode of *River Monsters.*

After an hour of fishing, our gillie suddenly disappeared, replaced by "The Incompetent Guy." That's how I thought of him as he messed around on his iPad. He also tended to disappear when I was hauling in fish. And because he wasn't doing his job, which is to get the other lines out of the water when I'm landing a fish, my lines were constantly getting tangled. Plus, he didn't understand that I was after gar, so he kept putting small bait on my line, and advising me to set the hook when a gator gar took the bait.

I was letting them take out line, however, and waiting for their second runs. But this wasn't Arkansas anymore, where river gar run hard and fast downstream, weighing their take in their teeth before deciding to commit. Since these gar were lake-locked, they'd just slowly move a few yards in any direction, lazily pulling the float behind. They were conditioned to eat small hunks of food, so they swallowed sooner. And with that weight dragging on the bottom, they were prone to drop what they scavenged due to the resistance. It was a tricky procedure to gauge their behavior.

Just like in the wild, where I only get an estimated ten percent of the gar that bite, I kept losing those alligator gar. Until later in the morning, the sky clearing up, I finally connected. That gator gar leapt and landed on its side, five feet long and heading for my net.

"Good," Robin said, already bored of hanging out in the torrid sun, waiting for me, waiting for gar. "When you get it in, can we go to some beaches?"

"Yeah," I said. "Just let me catch one gar, and I'll be satisfied."

That was the moment it got off. But as Robin groaned, I rolled with it. This happens to me all the time, and the more it happens, the more chance there is that I'll hit my window the next time. Or after that. Or after that.

Following lunch, the heat picked up and the fishing slowed. I saw a lot fewer gar, and they definitely bit a lot less. Occasionally, though, a float would shoot off or disappear completely, signaling a non-gar species. So I'd set the hook and pull in a ten-pound redtail, or a twenty-pound redtail, or a six-pound pacu, or a seven-pound tiger cat, which is a really cool stripy catfish with a cartoon-looking face.

All in all, I caught ten fish throughout that day, all the way up to twenty-three pounds. It was a blast to see all those species, and it was good for me to bait my own hooks when the Incompetent Guy slacked off. This allowed me to take control, cast where I wanted, and experiment more with letting the fish run with it, or just setting the hook—a technique I later tried a bit more, since anglers in these parts usually catch gator gar by accident, going for arapaima.

I even tried floating the bait near the surface by attaching a plastic pop bottle. But I just couldn't get a gar, even after four o'clock when the winds came in and cooled the lake. The burping catfish just kept on coming, and the gar kept eluding me at a maddening rate.

By the time our driver arrived to take us back to Jurassic Mountain, Robin was totally burned out from lack of shopping and sights to see, and I was totally sunburned, even though the day had been, for the most part, overcast. But as we wound through the pineapple fields, I was feeling optimistic—mostly because I was learning a lot about the first part of my mission: investigating this fishery.

"Ya know," I told her, "just because these fish have already been gathered and placed in a lake, that doesn't make it easier. Some people think these lakes are like pay ponds, where kids pull fish out by the pound. But as you can see, I fished all day and the gar beat me. So I'm being forced to re-examine my techniques."

Robin sat there with her arms crossed. I went on:

"There's a lot to learn from these fisheries, which are microcosms of the bigger ones. I mean, government-run fisheries, they have to stock fish too, and they have to deal with many of the things these fishing parks have to contend with. Like drought, die-offs, water chemistry, stuff like that."

"Okay," Robin said, turning toward me, "I'll play the Devil's advocate..."

Uh oh, I thought, already regretting opening my mouth. Because whenever she takes this role, I frequently end up revising my positions.

"You say these are fisheries," she said, "but are they really? I'm just thinking of the counter arguments you might run into..."

"Of course they're fisheries," I pretty much snapped, my blood beginning to steam. "They're commercial fisheries, but they reflect what's going on in the wild."

"How so?" Robin asked. "And to what end? I'm just saying..."

I could've argued, but the tension was rising. The last thing we needed was to get into a fight, so I dropped it. Granted, I knew I was reacting emotionally, but I also knew she was challenging me to challenge myself. Or maybe... maybe she was just cranky from having to watch me fish all day.

* * *

Two more carp were dead the next morning, and Jules had a lot on his mind. He'd been talking with David at Greenfield, who had the same problem. David said that this happens every year at this time when the weather gets still and hot and overcast. Just like in the United States, algae blooms flourish in the heat of the summer and deplete oxygen levels. This, so it seemed, combined with a parasite that usually rears its head in June, was making "the scourers" swim on their sides as they tried to rub off the irritant.

Jules was also apologetic that the co-owner of Greenfield Lake had ordered his most experienced gillie to weed-whack the property instead of work with me, and had stuck me with the Incompetent Guy. Jules told me he had talked to David, who was also apologetic, and had promised that when I came back the following day, I'd get to fish with Nook. Since Nook was the most experienced gillie they had, he was sure to get me my gar. As an incentive, Jules said he'd throw in a tip of 500 bahts (about seventeen bucks) if Nook could get me a gar, and I agreed to match that myself.

This day, though, would be all about landing a massive arapaima. I refused the pellets, chose two huge predator rods, and went out shooting for big game and big game only.

All day long, I sat there in the sun, where an eight-foot fatty had surfaced in the morning. I figured I would wait them out, that one had to swim by sometime that day. With fifteen behemoth arapaimas out there, my chances were pretty good that they, or one of the gator gars, would come by and take the bait.

Meanwhile, Jules called the government fisheries department and found out that a lot of carp were going down in the region. They requested a water sample, so his partner Eddy came over and filled a plastic water bottle, then drove it to the place.

It didn't take long for the report to come back confirming Jules' suspicions. The water was fine, but the stillness and humidity had caused some "stratification," which affected the chemistry of the lake and allowed for the growth of parasites. Which is just what my gillie had diagnosed, pointing at the water, then pretending to scratch himself all over.

By noon it was windless and hot as hell, I still hadn't gotten a bite, and the arapaima were everywhere. It was more than frustrating to watch them rise right above my bait, then swim on without a care.

Still, this lack of action allowed me to get some thinking done. The main thing I thought about was Robin's question about what these fisheries are good for. I didn't come up with anything really groundbreaking, but I did come up with three specific arguments.

First, like zoos or aquariums, micro-fisheries provide an educational service. People can watch fish and see how they behave, and learn from that, then apply that to the larger picture. For instance, Jules had noticed that anglers catch more arapaima three days before a full moon. There was also a way to spot their bubbles and figure out which way they were heading, then chuck the bait in front of them.

Secondly, our observations can serve the needs of the fish. For instance, if we see them dying from parasites in June (which corresponds to spawning season), then we know they need more protection at that time of year. Therefore, laws can be passed that are conscious of when fish are less stressed and in better condition to be reeled in, like we've done in Arkansas for alligator gar.

But my best point was this: Fishing parks take pressure off species living in the wild. For example, if a fisherman can travel to Thailand for a weekend knowing that he's going to a place stocked full of arapaima, he's liable to choose this over trying to locate a rare individual in the wild that he has a way bigger chance of missing. And if he does connect in the wild, he's more likely to hurt that fish than he is in this semi-controlled environment, which takes great pains to protect fish. And this, in turn, helps preserve apex predators in the wild—which provide balance.

In Thailand, arapaima fishing is an industry—one that hardly anyone ever heard of until the "Amazon Assassin" episode of *River Monsters* a few years back. After Jeremy Wade showed the world this furious ancient fighting fish, the demand was immediate.

That's why Peter from Holland came to Jurassic Mountain, and that's why I, in part, chose this lake: for a chance to land a truly unique freshwater monster. That's why arapaima are being farmed. That's why they can exist in the wild. Perhaps if we'd had fishing parks for gator gar in Arkansas in the fifties, they wouldn't have been extirpated in just three years, for which the entire fishery of the state has suffered—a simple formula being: the more big fish you have in a system, the healthier and more productive it is.

Still, it was mentally excruciating fishing twelve hours in the sizzling sun, not getting anything. Robin had gone off to the beaches of Hui Hin, so at least I didn't have to worry about her being bored. Technically, I should've been bored myself—but with those jagged mountains jutting up and those oxen-sized arapaima slapping down all around, I found it easy to just sit there doing nothing.

Nevertheless, by the end of the day, I found myself getting angry. Not because I wasn't getting what I'd paid for, but because I couldn't get what I wanted—which was dangling right in front of my face, just inches out of reach.

It's always unnerving for both angler and guide when there's a business transaction involving a fish and time is running out. In short, I'd missed my chance for a world-class arapaima, but I still had a chance for a smaller one at Greenfield. More importantly, Greenfield was full of the specific fish I'd traveled halfway around the world to get.

* * *

We arrived at 7:00 a.m., Jules and I, and went down to the shallow end, where David was "catapulting" chicken gizzards with a slingshot, chumming the area. It was cloudy out and there were arapaima and gar rising all over, plus a lot fewer pacu and catfish to contend with because of the depth. Both Jules and David were resolute that I'd get my gar today. Nook was also there, which was encouraging. There were plenty of pictures around the place featuring his smiling face on the tail end of numerous seven-foot arapaimas supported by at least two fishermen.

The question at that moment was why some gator gar have knobs on the ends of their noses and some don't. This question was raised by David's brother John Wilson, who's supposedly the most famous angling celebrity in the UK. John had visited the day before and had seen my previous gar book, which inspired this question.

My uncertain answer was that they all had that lump, but it was more pronounced on some and less pronounced on others. I also posited that alligator gar grow into their heads, so those knobs are huger when their heads are slimmer, and the broader their heads get, the less visible it becomes. But I also knew, from observing my hybrid for ten years, that scar tissue builds up in that spot when they bang their noses against the glass.

Anyway, we got two poles in the water, and they started hitting right away. Basically, you've got one second to decide what to do when a float jerks. Thanks to Nook knowing those fish, he could tell if it was just some punk messing around, or an arapaima, or a gar—so he'd quickly advise what to do. But before half an hour was up, I'd missed at least three big fish.

On the fourth one, Nook set the hook, and quickly handed the pole to me. It was a four-foot-long, forty-pound arapaima running out the drag. I fought it for about ten minutes, it didn't leap once, and then I horsed it toward shore where Nook was waiting with the net. It was a technicolored F2. Nook scooped it up, no problem, and everyone cheered.

Photo 43. Nook, Spitzer and Arapaima.

Photo Courtesy of Siam Fishing Tours.

Admittedly, I was kind of disappointed by its lack of spirit, but I was also glad to accomplish an objective. Some photos were taken, but the pressure was still on for gar.

Jules and David waited around for another half-hour while several gar ran with the bait, then dropped it before their second run. I caught a pesky twenty-pound redtail and threw it back without a picture. That sort of fish was just getting in the way.

Eventually, Jules and David took off, and Nook and I settled in under the shade of a large umbrella. The heat was picking up, the sky was clearing, and if the patterns of the first few days held true for today, this meant the fishing was slacking off.

That's when a gator gar hit, and slowly started tugging on the float. Rather than let it run this time, Nook yelled "Strike!" so I set the hook.

There was too much slack in the line, though, and since it didn't feel like I made the connection, I instinctively ran backwards until the rod bowed.

Something was on the line. Something that felt like a gar, since it came through the water with little resistance. And twenty seconds later, there it was, a three-and-a-half-footer coming in with its mouth agape. It was in my control, not even trying to defy the physics drawing it in, and it went right into the net.

This was exactly what I'd come to get. It must've weighed twenty pounds, and it looked just like any gator gar I'd ever caught in Texas or Arkansas. It had spots on the tail, a hardly distinguishable knob on the end of its nose, and its fins were red and chewed up.

Photo 44. Thai Gator Gar.

Photo Courtesy of Siam Fishing Tours.

I whooped, we took some pictures, and then I held it in the water until it regained its composure. It then snapped its tail and shot off, leaving me strangely satisfied. Since I'd accomplished what I'd set out to do, I spent the next five minutes shouting out stuff like "Yeah Baby!" and "That's What I'm Talking About!" Still, in the back of my mind, it felt a bit anticlimactic.

Whatever the case, the pressure was now completely off, and I didn't need to fish anymore. But since my ride wouldn't arrive for eight more hours, there was still a day of fishing left. So Nook and I, we kicked back, only making minimal conversation. He knew enough English to talk with me about tackle and the fish from where I came, but neither of us felt the need to do much more than sit in the shade in silence.

An hour later, a float suddenly disappeared, so I set that hook and felt the meat of a fifty-pound arapaima—one that was hardly resigned to letting me have the upper-hand. It ran like a deer and leapt like one too. When I got it a yard from shore, it shot straight into the sky and did a bit of tail-dancing, vibing out like a dynamo. But I brought it back, and Nook got it in the net, which this arapaima didn't accept. It shot skyward once again, taking the net with it. Nook held tight to the mesh, and the handle swung and cracked him in the knee, prompting him to employ a few English words that I hadn't heard any Thai person express yet.

Nevertheless, we landed it, and then Nook held it until it was strong enough to take off on its own—which allowed me to take a closer look. With that tiny muppet-looking head, it looked pretty funny, but the rest of it was a raging monster. All its fins were situated in the back, designed for the purpose of sheer torque.

Photo 45. Funny Face, Mongo Scales, Body Built for Running Down Prey.

Photo by Mark Spitzer.

And so the day went on, Nook and I sharing a strange intimacy. We were both sitting in the same shady spot, because he was of a different class. The other gillies I had worked with always slunk off to a nearby spot, then came over whenever I had something to deal with. Nook, however, was better at the language, and since he was used to working closely with the "great white hunters," it seemed appropriate for him to be by my side.

I had a notebook with me at all times, and one thing I wrote in it was "Feeling like Bwana." I mentioned this unsettling feeling to Robin later, who wasn't familiar with the term, so I explained its origin in the Tarzan books I'd read as a kid. Edgar Rice Burroughs had used that word to refer to the "white apes," which had a higher status than the other apes, who were always eager to serve.

I didn't feel like Bwana, though, after lunch. As the heat of the sun bore down, and as the wind stopped rippling across the surface, and as the afternoon activity lapsed, Nook and I both fell asleep—which I was okay with. But an hour later, when I asked Nook if he enjoyed his nap, he was reluctant to respond. He probably figured it was his duty to stay awake.

We had a beer, and kept sitting there, and nothing happened for hours. Then, around five o'clock, totally complacent in being patient as I had for the last four days, just sitting there staring at the trees and sky (which is why Robin was off exploring some jellyfish-laden beach in Cha Am), there was only one float to be seen.

I set the hook, expecting nothing, but was surprised to encounter enough weight to make the rod bow. I'd hooked another five-foot-long F2 arapaima, which fought back like a wildcat in a burlap sack. I didn't even need this one, so it was a bonus. A bonus I brought in and landed, adrenaline shooting through all my tubeways. The Incompetent Guy came over and snapped a picture.

Photo 46. Fifty-Pound, Five-Foot Arapaima.

Photo Courtesy of Siam Fishing Tours.

Two days later, we were back in Arkansas, where I was still reflecting on micro-fisheries and what they're good for. As I considered this question, I found myself dealing with my own fishery: a seventy-five-gallon tank with a hybrid gar and a bowfin in it. I'd let my spotted gar go two months ago to cut down on the pH and ammonia, which naturally build up in any aquarium. Lately, I've been taking a gallon out once a week to water houseplants, because that water is high in nitrogen. Then I add a gallon of dechlorinated tapwater back into the tank.

Since I've been living in the South, the infections have been coming on a regular basis. They're introduced by the minnows I get from the bait store, usually in the heat of the summer. You can never tell if a fish has fungus or body slime or popeye or tailrot or a thousand other illnesses until it's too late. It's always a pain to have to drain and treat an aquarium. I usually use a product called Triple Sulfa, which is pretty good at destroying harmful bacteria, but the process is expensive, time-intensive, and it always sucks to watch your fish fight off disease.

These days, I have special pumps in both my minnow tank and my gar tank, which disinfect with ultraviolet light. Since there's plant life floating around in the gar tank, it's always getting caught in the pump's filter and clogging it. Keeping it clean with a toothbrush is a daily activity, but for the most part, this diligence helps to keep my fishery relatively disease-free.

Other problems, of course, are always arising. My gar has a tendency to thrash around, sometimes busting under-gravel filter tubes. Then I have to get on the Internet and order the appropriate part. Filter stones need constant updating, and sometimes the water turns brown or green, and then I have to treat it with algae-eradicating chemicals, or add lake water with healthier micro-organisms.

But that's what you have to do if you want to keep fish. Fisheries equal maintenance. Sure, fish used to maintain themselves on their own, but now that people are part of the equation—adding and subtracting to the water quality and bio-mass of virtually all populations—the equation has irreversibly changed.

Take the cod trade for example. Nations were built upon this fish: Canada, the United States, Iceland, Greenland, Europe, et cetera. We couldn't have had the exploration and commerce we've had without the highly preservable nutrition source of "salt cod." Similarly, England's working class never would've had its main staple of fish 'n' chips without cod—a fishery that's been replaced by roughie, haddock, even dogfish shark. There have also been several wars fought over cod. All this is documented in Mark Kurlansky's "micro-history" *Cod,* published by Penguin in 1998. My point being: we thought the supply was unlimited, but it wasn't.

Now let's take a quantum leap, comparing fisheries to "Translation Theory," a subject I once studied at the PhD level in linguistics. The main thing I got from this study, which is applicable to other studies, is that even "bad translations" are useful, since they contribute to a dialogue in which the goal is to discover truth in one language so as to interpret it into another.

Along these lines, I'd say that all fisheries are useful (whether they be private or micro or government-managed), since they also contribute to a highly active dialogue—in which examples from one fishery can be applied to another, in which things we learn in one place can solve problems in another, in which the behaviors we observe in one situation can help explain those happening ten thousand miles from home.

In some fisheries you can fish, in others you can't. Some fisheries are sick, some aren't. But what they all have in common is a certain amount of meddling. In some places, the amount of meddling is close to zero. In other places, meddling happens more frequently. Fisheries exist because we exist, defining them as fisheries.

It follows then that if we didn't exist, fisheries wouldn't exist, because fisheries are man-made constructs. If we didn't exist, it would all be wilderness—but that's not going to happen, at least not in the near future.

In the meantime, here's my simplistic answer to Robin as well as myself, as to the value of sport fisheries. It won't satisfy everyone, but it's what I've got:

Ultimately, fisheries give us fish, and what fish are good for is stimulating the imagination—which might be even more important than maintaining balance in the wild. Because if we didn't have all that folklore and science and history out there revolving around the subject of fish, we'd have a lot less exaggeration in the world—and exaggeration makes the world a more colorful place.

That's why fishing parks exist in Thailand, making a colorful place even more vibrant. And that's why fisheries exist in the rest of the world, where cultures built on fish demand both commercial and recreational fishing for survival and escape.

So here's to fish and fisheries, and the challenge of catching the biggest, ugliest, fossil-fishes in our midst. Like gar and arapaima—which, through their reluctance to evolve, automatically inspire myth.

Chapter 10

Long Live the *Pejelagarto*!

A Culture of Aquaculture Thriving in Mexico

From what I'd read about the role of the tropical gar in the state of Tabasco, I knew I had to investigate the *pejelagarto* (meaning "fish-lizard"). Since there were numerous fish farms in Mexico raising gar for research and food, plus a culinary culture based on this Mexican tropical gar, I hopped a plane to Villahermosa, got myself a rental car, and took off for Comalcalco to meet Gabriel Márquez Couturier. "Gabo" runs the state-of-the-art Otot-Ibam aquaculture facility for gar, he's a biology professor at the Universidad Juárez Autónoma de Tabasco, and he'd been working with *pejelagartos* for thirty years.

Part of my mission was to find out why gar-farming had become so popular in this part of the world, and part of the answer was that the environment there had been severely crippled by the oil industry. For example, there'd been sixty-three leaks in 1993, ninety-nine in 1994, and 135 in 1995. According to a document entitled "Human Rights and Environment in Tabasco," written by an international human rights delegation in 1996, "The state of Tabasco on the Gulf Coast of Mexico is experiencing an ecological and political crisis stemming from more than two decades of intense and reckless exploitation of the state's petroleum reserves by Petroleos Mexicanos (PEMEX)," which sells 85 percent of its exports to the US. Among twenty types of damage to the ecosystem listed in this document are "hydrocarbon spillage" coming in at number one, and "toxic releases, hydrological disruption, and acid rain generated by petroleum extraction and processing... [which] has practically wiped out fish populations in many streams, rivers, and lagoons, effectively destroying the ancestral livelihoods of many Tabascans."

Seventeen years later, the situation hasn't improved. According to a May 16, 2013, article by the Inter Press Service News Agency entitled "Mexican Communities Sue PEMEX for Environmental Justice," gas flares in the municipality of Paraíso have mutated chromosomes in 10 percent of children in the area. Along the same lines, a 2010 study of earthworm populations published in *Research Journal of Environmental Sciences* (vol. 7, no. 1) states, "In the southeast of Mexico, the petrochemical industry activities have contaminated the environment, principally in the state of Tabasco, where oil spills have been reported frequently, with damage to fauna and flora." The article goes on to list commonly found contaminants in the soil such as phenathrene and anthracene, which "have been found to be toxic to fish," and the carcinogenic mutagen benzo(a)pyrene. Hence, fishing is toast along that coast and the alternative is aquaculture.

Wild-caught gar, however, are still a commodity—which I saw as soon as I got lost in the marshy lowlands. There was a guy selling gar by the side of the road. He had a stringer of four and a family in a car had just bought two. By the time I parked on the shoulder and jumped out with my camera, there were only two *pejelagartos* left. The fisherman let me take a picture.

Photo 47. Garmonger between Villahermosa and Frontera, Tabasco, Mexico.

Photo by Mark Spitzer.

Two hours later, after asking directions at every other crossroad, I made it to the city center of Comalcalco and found my hotel. Since my Mapquest directions hadn't done squat for me, I took a taxi to Otot-Ibam, arriving only a few hours late—and without the translator I had promised, due to an emergency. Gabo's limited English and my abysmal Spanish would therefore have to do.

I was welcomed warmly by Gabo, who was just wrapping up a birthday party for one of his workers. Cameras came out, and suddenly I was posing with the birthday girl, because in the gar communities of Tabasco, I'm a semi-celebrity. Having been on the alligator gar episode of *River Monsters*, and being in some of the top Google images for "gar," this was just the beginning of a day packed full of posing for photos with gar farmers, gar researchers, gar workers, gar businessmen, all sorts of gar people. Gabo later joked that he should charge twenty pesos per photograph.

But back to Otot-Ibam, where Gabo and I exchanged books. He had co-written a 2013 study entitled *Acuiculturea Tropical Sustentable* with César Jesús Vázquez Navarrete, Wilfrido Miguel Contreras Sánchez, and Carlos Alfonso Álvarez González. This book was written in four different languages, and if it's not the bible yet on how to raise and care for tropical gar, it will be soon.

Gabo then showed me around. The grounds were full of ponds and tanks and vats. The first one we came to was a concrete spawning tank the size and shape of a standard, suburban, above-ground swimming pool. This is where hormone-induced females jettisoned eggs onto green plastic plants. Three to nine males per female add sperm (even in the wild there are more males than females), and the tank then serves as an incubation chamber.

The next giant tank held a roiling host of diverse gar, varying in size and color. In addition to the regular gray *pejelagartos* there were "blancos" and "black panthers," which are very rare in the wild (.025% of the population), but are now being raised by the thousands in Tabasco. There was even a strain with spotty reddish markings called "jaguar."

Photo 48. A Blanco.

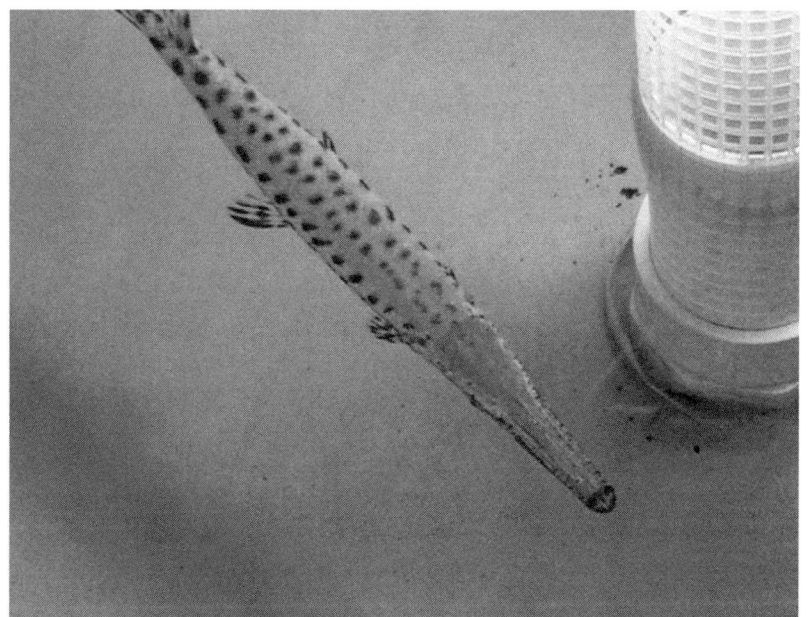

Photo 49. A Tub Full of Black Panthers.

Photos by Mark Spitzer

After that, we toured the juveniles, which were being kept in what looked like oxygenated washbasins. There were all sorts of sizes, ranging from less than an inch to nearly three inches. There were small round pellets floating on the surface, which the gar devoured, no problem.

Then we visited the adults, which were held in hot-tub-sized containers being scrubbed by diligent workers. Many of these tanks held ripe females, which I incorrectly called "Gorditos," meaning *fatsos*. Gabo corrected me, ending the word with the female "-as."

"Come on, Mark," Gabo said with a wave of his hand, and led me to the largest tank, which had been roiling with gar when we entered the farm. It had been about four-feet deep, but now it was down to a few inches, gar backs squirming all over the place.

Gabo jumped in and motioned for me to follow, and I took off my shoes and followed him. So as not to startle the gar, we squatted down close to the fish, which were swimming and swishing all around us. Basically, we were rustling some up for a photo op, and I paid special attention to how Gabo picked them up. Moving slowly, so as not to stress them out, he'd find one he wanted, gently place his hand beside it, then gradually move his palm underneath. If it began to freak out, he'd let it get away, but if it didn't, he'd roll it over on its back. Then lifting it upside-down, he'd cradle it like a bambino. I followed suit. Apparently, this method calmed them by making their eyes point down, so they didn't have to see the alien faces of us humans.

After that, the question was "*Cerveza?*"

I responded with an enthusiastic "*Sí!*"

Photo 50. Gabo y Pejelagarto.

Photo by Mark Spitzer.

In the truck I asked Gabo about the Mexican adopt-a-gar program, which I'd read about in a 2003 newsletter put out by Oregon State. The article, entitled "Outreach Program in Mexico Captivates Schoolchildren," stated that "Wild gar populations have been severely affected by the absence of catch regulations and the loss of degradation of spawning and nursery grounds." It then explained how participants in the program receive two juveniles in a cup and a month's supply of fish food. The kids then raise their gar and reintroduce them into protected habitats. In the process, the children learn about "the gar's role in wetland ecology, its life history, evolution, distribution, and fishery situation." Gabo said the program was still active, and that the children receive their *pejelagartos* in March.

Prior to Mexico, I tried to get to Cuba, where they have a similar program centered around the *Manjuari,* or Cuban gar. I had hoped to tour the aquaculture facilities in Zapata National Park, where Dr. Andrés Hurtado was doing highly innovative work with this fish. I wasn't able to get my research visa, though, due to poor communication, the sudden death of an administrative contact, invasive government oversight (in Cuba), and ultimately, the US Government Shutdown in 2013, which shut down my research license application. Sneaking in not being an option, I'd opted for Mexico instead, which actually had more gar to offer because of the *grande* gastronomical culture.

Coincidentally, Gabo told me that he had studied for his masters degree in Havana with his good friend Dr. Hurtado. Our communication wasn't good enough for me to determine if Gabo had studied *under* Dr. Hurtado, but since both Mexico and Cuba have adopt-a-gar programs for school children, it seemed obvious that both Gabo and Hurtado shared some ambitious visions about stocking gar in the wild.

Gabo told me that gar farming in Tabasco served three main areas: conservation, food, and the ornamental fish industry. I wasn't ready to get into the dynamics of the latter, but I was extremely interested in the first. Gabo reaffirmed how petroleum contamination had diminished the fisheries, and he told me that he knew of thirty-two gar farms in the area, which produced 400 to 600 tons of *pejelagarto* per year. He also told me that there are no alligator gar in Tabasco, that tropical gar has a protein value that's similar to salmon, that gar meat is a major source of nutrition for Tabascans, and it's considered an aphrodisiac.

* * *

We drove through the palm trees and arrived at a fish farm owned by Gabo's friend Martin. Gabo showed me around. The tanks were square and concrete, like esophagi. Martin raised gar and tilapia, plus goldfish that he fed to his gar. He also had a couple of ponds dug into his property, one of which Gabo said had two thousand *pejelagartos* in it.

Gabo introduced me to Martin and his crew, more photos were taken, and then Gabo showed me Martin's juveniles. Again, there were a lot of white and black gar, which I assumed were meant for the ornamental fish trade. I asked Gabo about this, and he told me they were, adding that when they're three years old they sell for $5000 a piece. The buyers, he added, were mostly Japanese, English, German, and Chinese. The females, however, were not for sale. Gabo and Martin, being shrewd businessmen, only sold males, so they couldn't reproduce.

We continued on to Miguel Garcia's lab tucked amidst the coconut trees. Miguel came over and fist-bumped Gabo, and they took me inside. There were tubs full of fingerlings and larger tanks full of larger gar, though not many over a foot long.

Driving to the next spot, I asked Gabo why most of those *pejelagartos* were small.

"Cannibalism," he said, and explained that they were being raised to feed to larger gar.

Then we went to the university in Villahermosa where Gabo works. He introduced me to a researcher named Lenin Arias, who spoke English and showed me their *pejelagartos.* Big vats, big specimens, and big ideas too.

"We will clone these," Lenin told me, indicating a tank full of multi-colored gars. Then he said the word "androgenetics," followed by "gyno-genetics."

"We are creating a genomic bank," he said, "to preserve all types of gar."

This led to a conversation about how some populations in Chiapas have different genetics from those in Tabasco. I asked about Guatemala, where another isolated population was also considered a potential different species, and Lenin said that their objective is to ensure that if a gar specie crashes they can reintroduce it with the correct endemic genetics, rather than those from some other place.

Then the word "triploids" came up and Lenin showed me some four-to six-inch *pejelagartos*. He said that they were three years old and had been bred to stay small so they don't outgrow their aquariums. He also told me that they were sterile.

I joked to Lenin that the work he was doing was "avant-gar," and he laughed at that. Still, I was pretty serious, and still, I was ready for that beer.

* * *

Which turned into three at La Palapa de Polito, an outdoor gar eatery and watering hole on the outskirts of Villahermosa—where pictures were taken with Polito (the restaurant owner and Gabo's business partner), Polito's girlfriend, a gar fishery manager named Mary, an attorney/gar farmer named Roberto, Gabo (or "Maestro," as his friends called him), and myself.

The gar then arrived in two forms. First as *salpicón de pejelagarto*, which was basically pulled gar meat mixed with purple onions and cilantro, to be eaten with tortilla chips. And it was damn good, especially with some lime squeezed over it and a splash of chili sauce.

Then we had smoked gar on a stick. It was a roasted fourteen-incher that had been gutted and blackened on both sides with its scales on, its head on, and mouth gaping open. Polito brought it to the table overflowing from a platter, inserted a fork at the base of the dorsal fin, then worked it like a can opener toward its head, splitting the hide open. He peeled the shell away from both sides and began breaking up the meat. It came right off the bones in two big flanks, leaving the back strap above the spine.

Photo 51. Pejelagartos at Palapa de Polito.

Photo 52. Pejelagartos at Palapa de Polito.

Photos by Mark Spitzer.

Gabo showed me the traditional way to eat *pejelagarto asado*. You break off a piece of thick corn tortilla with slivers of roasted garlic on it, then get yourself a hunk of gar, which you break up on the tortilla. Then you sprinkle some sea salt on that, add some lime juice and chili sauce, and put it in your mouth.

The flavor was only slightly smoky. It was delicate and had a dense, dry texture, a lot like chicken breast. It was whiter and lighter, though, with a very mild non-fishy flavor. It was excellento!

At one point I noticed some large tanks behind the restaurant and decided to take a look. They were full of tilapia, as Polito pointed out, coming up beside me. Then he motioned for me to follow him, and he took me over to a pond swarming with tropical gar, all of them eating size. There were hundreds in there.

I saw the logic. Not only was this place a roadside restaurant/bar, it was also a gar farm, where Polito could subtract as many fresh gar as needed per day, then roast them for his customers. It was sustainability in action.

Polito went over to a plastic sack and started throwing handfuls of trout pellets out. Having been raised on these pellets (which are over 42 percent protein) as well as live fish, the gar came swimming over for their supper, breaking the surface and gobbling like crazy. Since I'd never seen gar feed on anything other than live food, this was a bit weird to watch. Tropical gar, however, are part of the "broadhead" family of gar, which includes alligator and Cuban gar. And since gator gar are scavengers that pick hunks of meat off the bottom, I figured that the broadheads, in comparison to the "slender gars" (that's the other designation) were more prone to eat non-live food. Since longnose, shortnose, spotted, and Florida gar preferred live food, I figured they might be harder to farm. But you never know. I've heard of Florida gar eating pellets, and there are YouTube videos of spotteds eating them as well.

Gar feeding habits, of course, depend on what they're conditioned to consume. Solomon David at the Shedd Aquarium has since informed me that all gar species will eat pellets if trained to do so.

Back at the table, I looked into the jaws of the tropical gar we'd devoured and saw its double row of teeth on the upper jaw. This is also a characteristic of the alligator gar. I asked Gabo if this held true for the Cuban gar as well, and he said it was a trait of the broadheads.

About two years ago an article came out in the *Archives of Natural History* written by a biologist in Winona, Minnesota, taking me to task for claiming that alligator gar once ranged as far north as Canada. My argument was based on the journals of Samuel de Champlain, who, in the early 1600s, reported that Native Americans in the Lake Champlain area had shown him the skull of a big gar called "*chaousarou*," and that there were two rows of teeth on its upper jaw. My respected colleague countered that this fish was a longnose, which have an outer-row of small teeth and an inner-row of larger teeth on their top jaw. The double row is a lot less prominent on the longnose, but my colleague didn't note this. Instead, he very cleverly accused me of being guilty of distorting information—but here's the deal: if you look into a broadhead's mouth, you see two distinct separate rows of teeth on the upper jaw, and if you look into a longnose's mouth, it looks like the same row of teeth.

Meanwhile, we kept drinking some sort of local aperitif made with anise. They started out small and got larger as the night went on, until finally we were stumbling through another gar farm, getting bit by the mosquitoes, and posing for more "*monstruo de río*" pictures.

On the way back to my hotel, Gabo stressed the importance of aqua-culture as a method for keeping *pejelagartos* in the "deteriorating" system to help control other species. This is an idea I could relate to, because that's what happened in Arkansas. When alligator gar were extirpated, populations of destructive fish grew stronger, resulting in a weakened system.

But then I thought of another species, which aquaculture was also serving: the Tabascans, whose gastro-culture, based on the tropical gar, would surely go extinct if it weren't for fish farms preserving their natural heritage.

Of course, there's no way to measure what's more important: gar as a nutritional, low-cost food source, or gar as a cultural component that you can't just replace with chicken. If you did, a lot of colorful, historical context would be lost. The point being: Sustaining people through sustaining gar is vital to the future of Tabasco. That's why the independent financial institution Fondo para el Medio Ambiente Mundial (FMAM) promotes subsidies for *pejelagarto* research in Mexico.

Still, the gravity of gar in Tabasco wasn't completely apparent to me. I knew it was there, I knew it was real, and I knew it meant more than just serving as a source of protein, but as of yet, I hadn't seen enough of the relationship between this fish and its people to get a handle on the cultural context of *pejelagarto* in Mexico.

* * *

In the best-selling book *Four Fish: The Future of the Last Wild Food* (Penguin, 2010), Paul Greenberg uses a benchmark formula that's sometimes employed to consider species for domestication. He writes, "Irrespective of his infamous eugenics writings, anthropologists consider the list of criteria set out by the nineteenth-century intellectual Francis Galton as being a good thumbnail sketch for what guided Neolithic humans" to farm certain animals. Greenberg then demonstrates how the sea bass fails, even though this fish has become a highly popular farmed food source.

So let's plug tropical gar into Galton's equation and see what happens:

1) They should be hardy: Greenberg relies on the argument that a good candidate in this category lays a lot of eggs, has a high survival rate, and is resistant to disease. Gar, of course, lay tens of thousands of eggs in one shot, and since they do well in low water, warm water, water lacking oxygen, and can grow fast enough to eat in just one year, they are hardy. Also, since farmed *pejelagarto* larvae have an 84 percent survival rate (this figure comes from Gabo's book), gar are definitely more hardy than sea bass, which (according to Greenberg's book) have a 99 percent die-off rate. As for disease, Gabo tells me, "There is no contamination" in aquaculture facilities. Simply put, diseases that naturally occur in the wild do not infect gar farms, where quarantine tanks are actually employed for forty days. Even common parasites like *argulus* (sea lice) get weeded out. And because water is constantly being changed, fungus, ick, body slime, all those infections, they don't matter.

Thus, gar are tough, gar can endure, and they satisfy this criterion.

2) They should have an inborn liking for man: Gar don't hate humans. If they did, they'd attack us and try to intimidate us, but that's not in their nature. If anything, gar get along with us. From being frequently handled in aquaculture environments, which is part of the philosophy of farming gar, *pejelagartos* don't object when humans wade among them, reach in their tanks, and move them around. When Gabo and I were in that tank, they swam right up to us, passed between our legs, and always had smiles on their faces. When Polito fed his stock, they came swimming over gleefully, dancing for their dinner. Basically, as long as we move slowly, so as not to startle them, they like us enough to allow us to touch them. In fact, I've been petting my pet hybrid for years and he doesn't mind at all.

So score one for the gar.

3) They should be comfort-loving: This category refers to whether a species can deal with containment or not. Some fish don't do so well. In *Eels: An Exploration, from New Zealand to the Sargasso, of the World's Most Mysterious Fish* (Harper Perennial, 2011), James Prosek writes about how these fish sometimes commit suicide by slamming their heads against their tanks. Gar, however, adapt well to small spaces. I've been watching my hybrid dance in front of his reflection for years. He even taught a spotted gar to do the same. For years, they kept themselves entertained by going up and down and moving around and stalking minnows when they weren't even hungry. So they're good at maneuvering, hovering, turning around, and swimming backwards to while away the time.

Gar, therefore, make themselves comfortable.

4) They should breed freely: This is another area in which gar do well. In aquaculture facilities, the females grow fat with roe, and then they're moved to the spawning tanks. As Gabo and co. explain in their book, "A rectangular 4x8x1 m [tank] is ideal for . . . spawning. Nests can be made with tufts of grass anchored to the bottom simulating a flooded plain . . . Water depth must be from 60 to 80 cm. . . . . three 3 to 4 kg females and nine 0.750 kg males . . . are set in the morning and 24 to 72 hours later a spontaneous spawn should occur with an estimated production of 25,000 to 90,000 eggs, depending on how many females spawn."

If that ain't breeding freely, I don't know what is.

5) They should be easy to tend: This is something I can attest to myself. If the possibility of contamination is lowered to nada, as Gabo says is the norm for farmed gar, then all you have to do is feed your stock, keep their tanks clean, protect them from predators, and add new water once in a while. As for newborns, they come with their own food supplies, which they live off until they develop out of the larval stage. Feeding with pellets it also low maintenance and cost-effective.

Gar pass this test with flying colors.

The verdict being: Since tropical gar score well in five out of five criteria, it's reasonable to assume that more sophisticated methods of evaluating "farmability" would produce similar findings.

As for the eugenics question, one can't help wondering what would happen to humans who don't score as well as gar on the Galton test.

Photo 53. Market Gar, Comalcalco, Mexico.

Photo by Mark Spitzer

There were plenty of gar at the central marketplace in Comalcalco, and they looked pretty healthy, considering reports of their toxic environments. Brought in by local fishermen, these gar had been caught in the wild.

As I wandered amidst the fish stands, I also saw tilapia, snook, mahi-mahi-looking fish, and mongo catfish. Some of those cats must've weighed eighty pounds. But there were also tarpon with their heads cut off, lying there in hundred-pound slaps.

Which brings me to my secondary mission in Mexico: to catch myself a tarpon. They'd eluded me in Nicaragua and had burned their impression into my mind by leaping ten feet into the air, shuddering silvery in the sky, fighting like no fish I'd ever had on the line, then leaving me with nothing but sore arms and a score to settle.

So eight hours later, after slaloming my micro-Hyundai rental car through a freeway cratered with potholes, I arrived in Champotón, a fishing town in the state of Champeche. I'd secured the guide service Tucan Sport Fishing, and Captain Edward would be taking me out the next morning.

But the result was a total bust. We fished all day on the Rio Champotón and I missed eleven fish. I finally hooked a tarpon, though, and fought it for a few seconds before it snapped the line. Still, I didn't give a crap. I enjoyed Edward's company, and having fought this fish again, my determination to get one was now even stronger.

Then, on the way back, I got stopped at a police checkpoint where a machine-gun-toting military bully tried to shake me down for a bribe. The same guy had done this two days before on my way up the coast, and just as before, I played the ignorant touristo, repeating "*no español*" and "*no comprender*" until finally he let me go.

But before I could take off, a jolly English-speaking hombre asked if he could catch a ride to Villahermosa. *What the hell?* I thought, and motioned for him to jump in.

Jesús, it turned out, was a chef and had cooked plenty of gar in his time. He described all sorts of recipes and ways to prepare gar as we passed local fishmongers on the road displaying stringers of *pejelagarto*.

I'd been planning on eating one last gar-meal in Tabasco before flying back the next day, in order to get more of a feel for the local cuisine. So after dropping Jesús off at a bus stop, I went to the airport to return the rental car.

* * *

That's when it started sinking in.

In the airport giftshop they were selling ceramic sculptures of *pejela-gartos* emblazoned with the word "Tabasco." They were also selling other various brick-a-brack with the tropical gar proudly displayed.

I took a taxi to Centro and passed a lot of food stands on the way, many of them sporting hand-painted tropical gar, advertising the local staple. Los Tulipanes, however, was a bit more upscale than the restaurants I'd seen out the window. This one had been recommended on various websites for their *pejelagarto empanadas*, so that's what I ordered. *Ensalada de pejelagarto* and *pejelagarto asada* were also on the menu.

My empanadas arrived all golden and fried with a bowl of pico de gallo. I bit into one and looked inside. The meat resembled shredded chicken, but it was orangish, as if it had been marinated in a peppery tomato sauce. Again, it had a very mild, unfishy flavor, though subtly spicy, and not dry at all. I added the salsa, then lime juice and chili sauce. They were juicy and delicioso!

I'd done some research on the Internet and had found tons of *pejelagarto* being served in the city. Some places had *pejelagarto ceviche*, others had *moles*, and others gar tostadas. Sites like *Travel Advisor* and *Foursquare* had shown maps of Villahermosa with all the hits for restaurants that served gar marked by flags. Needless to say, those flags were all over the place.

For a few minutes, I regretted going up the coast and getting skunked when I could've spent two whole days in Villahermosa touring restaurants and eating gar. But then I went for a walk, searching for the city center, and en route I saw mucho *pejelagarto*.

On a bridge crossing the river, there were vendors hawking actual shellacked tropical gars affixed to bases proclaiming "Tabasco, Mexico." I also saw many hand-carved pen holders and folk-art gar key rings. I bought a few trinkets and continued on.

Photo 54. Taxidermied Pejelagarto.

Photo 55. Pejelagarto Pen Holder.

Photo 56. Pejelagarto Key Ring.

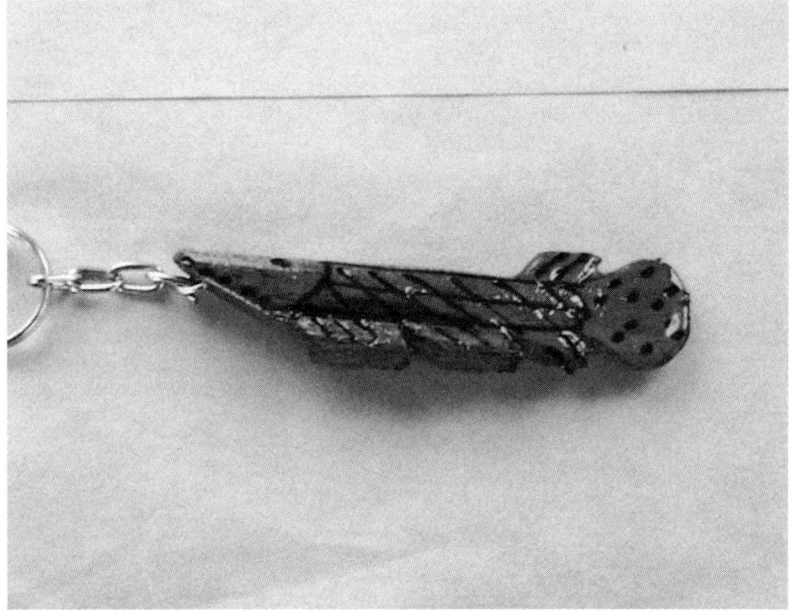

Photos by Mark Spitzer.

In the distance, I could see a palace-looking place flying giant Mexican flags, so I figured that might be the square. I was correct, and the closer I got, the more gar stuff I saw. There were tables set up in the square, where people were selling *pejelagarto* T-shirts and *pejelagarto* baseball caps. And the souvenir shops around the square, they were selling *pejelagarto* plaques, clocks, plates, platters, candles, everything *pejelagarto*.

But it wasn't just the capital city of Villahermosa associating itself with this fish. Frontera, to the South, was also well known for its gar. Jesús the chef had pointed out signs for various cities on our ride that were famous for gar. Because *pejelagarto*—it wasn't just a symbolic fish that sold well as tourist fare—it was the icon of this place and its people. For centuries, it had been that way. Tabasco was loco for *pejelagarto*.

Then, the next day, at another airport gift shop, I saw *pejelagarto* napkin holders, *pejelagarto* key hangers, *pejelagartos* cut from stone, and even a cartoony stuffed *pejelagarto* with a goofy sombrero. I took some pictures of this garaphanalia, then flew off to Houston, wearing my new *pejelagarto* shirt.

Photo 57. Stuffed Animal Gar.

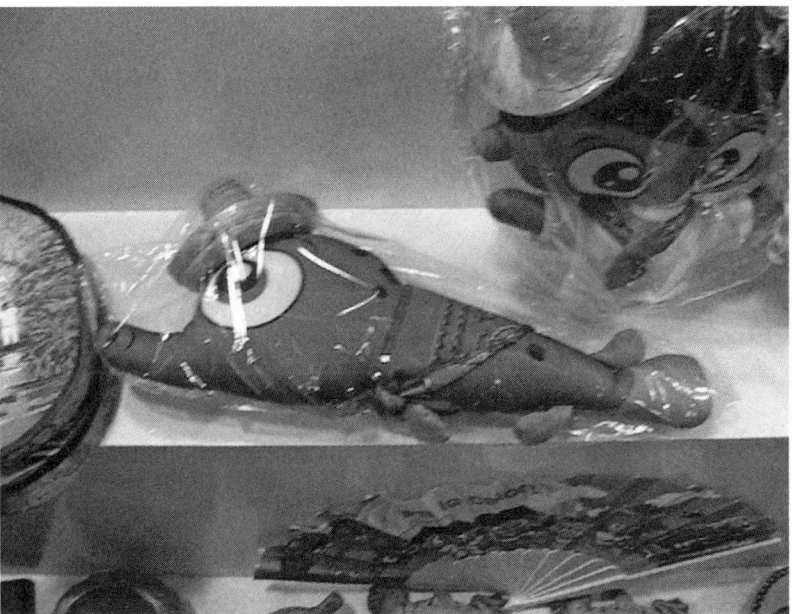

Photo by Mark Spitzer.

More importantly, I now had a better understanding of the cultural context of *pejelagarto*, my revelation being: Tropical gar aren't just an ingredient that goes into tacos in Tabasco. If anything, *pejelagartos* are Tabasco. They're its culture, its past, its future, and identity—which environmental destruction can't even take away. In fact, gar are as much a part of the people of Tabasco as the eagle is symbolic of the American spirit.

Of course, we brought the bald eagle back from the brink of extinction in the United States. Whether Mexico can do the same for the wild *pejelagarto* remains to be seen.

In the meantime, aquaculture projects in Tabasco are not only doing impressive work to preserve the *pejelagarto* and its culture, they're also providing a highly successful example of how to make use of an incredibly abundant natural resource—one that's generally considered "worthless" to the North. So let's meditate on that a bit, and ask ourselves if we've got what it takes in the US to re-envision the stigmas we ascribe to gar. After all, we are capitalists—so why not profit from sustainable gar-farming?

Conclusion

Return of the Gar

Without any warning, the sixty-five-year-old Pegasus pipeline burst in Mayflower, Arkansas. It was a twenty-two-foot rupture in a twenty-inch-diameter hose originally designed to pump unrefined oil from the South to the North, where the refineries used to be. Now, however, it was pumping a rougher viscosity Canadian tar-sand "heavy crude" in the opposite direction. The result was, according to the Pipeline and Hazardous Materials Safety Administration faction of the Department of Transportation: an estimated 3500 to 5000 barrels (or 200,000+ gallons) of toxic, noxious, fuming "bitumen" (as ExxonMobil calls it) burbling up from a suburban lawn.

It immediately filled the streets, flowed into rain gutters, found its way into a drainage creek, and flowed straight down the hill. Twenty homes had to be evacuated due to the poisonous air that sickened children in the school by the spill. But humans were the least of my concern. That crap was spilling straight into Lake Conway, on which I live.

Suddenly, we had a disaster on our hands, one of the hugest inland oil spills in American history. Trucks began to rumble in and guys with hazmat suits commandeered the creek, spraying undisclosed chemicals.

I saw this with my own eyes, as well as the armies of workers and checkpoint cops and road blocks and paper towels—which brought national attention from comedian Stephen Colbert, who made fun of the ridiculous fact that ExxonMobil actually used Bounty, the "quicker picker upper," to soak up hundreds of square feet of polluted marshlands.

Booms went out into the cove. Row after row, to stop the gunk from getting into the main lake, whose only defense was one mere levee with an open culvert, where thirty toxic chemicals—including benzene, toluene, ethylbenzene, n-hexane and xylyene—were spewing out. A boom was placed on the other side, but that didn't stop all that toxic water from flowing straight into the largest manmade fishing lake in the state as thunderstorms beat down for days.

Photo 58. Live Screenshot Displayed on *Arkansas Times* Blog, April 11, 2013.

There was a debate as to whether oil was in the lake, the fist-pumping, drill-baby-drill faction of the local petro-patriotic front sounding off against the more environmentally minded citizenship, even though Attorney General Dustin McDaniels clearly stated that if there's oil in the cove, there's oil in the lake. Meanwhile, rescue crews pulled out ducks, nutria rats, snakes, turtles, and all sorts of species slathered by disaster. Some were dead, some were alive, as the eco-anarchists came flooding in, organizing those in the "sacrifice zone" by handing out info on safety and health.

ExxonMobil tried to cover up their mess by bullying the authorities into creating a no-fly zone over the spill so that new crews couldn't broadcast scenes of apocalypse. A snazzy PR campaign was then launched that was reminiscent of the still-sore BP catastrophe in the Gulf, ExxonMobil proudly proclaiming, "WE'LL BE HERE UNTIL THE JOB IS DONE!"

I spoke out once at a town-hall-type meeting, urging action on investigating what those guys were spraying in the creek. During the BP debacle, "chemical dispersants" had been used to break up the spreading slick, and according to multiple sources, those chemicals were just as toxic as the spill itself. But for the most part, I was lazy.

Sure, I took my ecopoetics class down to the spill to see it firsthand: the clean-up crews, the destroyed vegetation, the private security guards telling us to get off public property. I even took a fishing pole with a paper towel rubber-banded to a lead weight to pick toxic sludge off the bottom by the dam, where there'd been a popular social media report of a fisherman pulling in a crude-coated lure—downstream of which, my friend Scotty caught a cast-net full of tar-sand gumbo. We could smell the fumes in the air, we could see the destruction all around, and my students were enraged.

People kept asking me to write letters to the editors in order to put this travesty into perspective, but I was too busy being caught up in the business of everyday living: teaching, writing, researching gar. Meaning

I didn't fight for my lake, I didn't fight for my fish, I didn't do enough. And for that, I had to accept the uncertainty that comes with non-action.

There were no figures for benzene levels or overall water quality. The state was withholding their test results until after ExxonMobil made their own report, and other agencies were not disclosing their findings as well. I tried for months, but just couldn't get a clear answer about whether the catfish and bass were safe to eat.

So I was having a conflict running my trotline, which was what I'd been doing for the last six years, and what I looked forward to twice a day. This was my reason for living on the lake, for communing with nature on a daily basis, for getting routine exercise paddling my canoe. I loved going out in those cypresses, seeing herons, ospreys, eagles, egrets, pelicans, turtles, and leaping shad. But with that carcinogenic goulash in there, I just couldn't bring myself to bring home the fish-bacon, which I couldn't bring myself to eat. So all I was doing was catch and release and damaging eyeballs in the process. Other times, though, fish would die on my line, or I'd snag the occasional cormorant, just so I could get my fix of the lake I loved.

This led to a serious conversation with myself, resulting in the decision that my relationship with the lake had run its course. I just couldn't continue to fish on it. It had been defiled. So I quit with it.

And I'm glad I did, considering the October 2013 data assessment report that finally came out on the Arkansas Department of Environmental Quality website. This report, which has since been deleted from the ADEQ menu on reports regarding the Mayflower oil spill, was written for ExxonMobil by a private contracting firm and was filled with misleading language. For example, "Based on the screening results, concentrations in 50 of the 54 samples were at levels that do not warrant further evaluation." Such spins make the figures look acceptable, but the fact remains that four samples were at levels that warrant further evaluation.

Other instances of spin from Big Oil can be found in the report's final analysis for surface water samples collected between March 29 and September 6, 2013. Eg, "only three of 70 VOC's (benzene, isopropy-lbenzene, and total xylenes) have been detected at concentrations above ESVs" [ecological screening value]. Since "only" indicates a judgmental value, it's hard to view this report as objective, so therefore honest. Many of the figures provided are specifically for the contaminated cove, and it's often reported that the main lake does not contain such concentrations. But given the biased voice of the report, I wasn't about to trust ExxonMobil. Since the cove is part of the lake, and since what comes out of that cove gets into fish, and since "Only four of 18 PAHs (anthracene, benzo(a)anthracene, benzo(a)pyrene, and pyrene) have been detected at concentrations above their respective ESV's in . . . Lake Conway surface water samples," the report still can't shirk the fact that hazardous levels were detected and reported in the main lake. This same page of the report also lists "Only two of 12 metals" detected in Lake Conway surface water samples "at concentrations above ESVs." The upshot being: Despite numerous reports about fish in Lake Conway being safe to eat (most of which came out before the data assessment report), that water, at the time of this writing, is an ethyl-methyl benzene stew of dangerous toxins swilling with carcinogens. So if you take a chance based on your faith in what fishermen say in chat-room conversations, you're taking a chance not only on your own health, but to whomever else you feed those fish.

It's unconscionable, of course, that state and federal authorities are not warning people about the cancerous chemistry of Lake Conway, but that's the way it is. People would freak out and flock away if the newspapers published such news, which tends to get banished to books like this, where the impact is less direct.

To be fair, I later wrote to Professor Jennifer L. Bouldin, Director of the Ecotoxicology Research Facility at Arkansas State University, who had been commissioned by the local paper to study the toxin levels in the lake. Her team had given the lake a clean bill of health, but I had to be sure. She replied, "The work we performed did not show any direct toxicity to the test organisms. Only sublethal (subtle) effects . . . To the lake's advantage is the cleansing power that water has—especially flowing water—even though this is not a flowing system, the streams that lead into and out of the lake are. Microbial action will diminish the harmful contaminants over time."

This is the philosophy often referred to as "the solution to pollution is dilution," but what it comes down to is this: Would you feel comfortable eating those fish or feeding them to your family? I put that question to Professor Bouldin, but didn't receive a response to that—though I'm certain of my own response. I'm not about to take that chance until we know how much time it takes to diminish those "harmful contaminants." Would you?

Photo 59. Hybrid Alligator-Longnose Gar.

Photo by Mark Spitzer.

This toxicity, unfortunately, became a metaphor for a point I'd reached in my life. With the warmer weather, the contaminations were becoming more common. The minnows I was feeding my hybrid gar and bowfin had fungus and other diseases that even the ultraviolet filter lights in my tanks couldn't fry away. When my bowfin went belly up, I knew I had a new lethal slime in the mix, and my gar would be going down next. So as I treated that tank and changed the water and did all that stuff I always have to do when an aquarium gets infected, I considered letting my gar go.

After all, I'd had it ten years, and after this illness was vanquished there'd be more to follow. Technically, my gar should've been at least six feet long, but its growth had been stunted by a once-a-day feeding schedule that I'd been robotically following for a decade. I figured that in freeing him I'd free myself from all that Mobius maintenance. But ultimately, I wanted my gar to be free, and to experience all the joys and terrors that come with that—just like me. Besides, what else could I learn from observing him that I hadn't learned in the past ten years? It was time to let him go.

When his health returned, I put him in a blue plastic tub and drove him to the Arkansas River. It was an emotional thing for me to do, but as I kept reminding myself, this wasn't a dog, this wasn't a cat, it had no human feelings. Plus, what I was feeling was only nostalgia, something it couldn't feel for me.

Deeper down, though, I knew that what I was feeling was way more profound than just letting go of something I loved. I was trying to free something in myself, something that had become routine, something that I needed to change in order to grow as a human being.

But when I lowered my gar into the river, he hovered there in my hands, not taking off. It was almost as if he didn't want to leave the security he'd known being enclosed in tanks all his life. Then he started motoring off, slowly and cautiously. For the first time in his existence, there wasn't a glass wall two feet from his nose, and I think this frightened him. There were sticks beneath him, rocks, current. He swam a foot and then another. Then three feet, four feet, five feet, still swimming on the surface.

As he reached the six-foot mark, the seven-foot mark, never looking back, his coppery color suddenly struck me. He'd taken on this metallic glimmer from living in a sodium-sulfa-treated tank for the last week, and his black tiger stripes were highly defined. Without question, he was the coolest-looking gar I'd ever seen, and just as quickly as this realization hit me, he was gone.

Photo 60. Return of the Gar.

Photo by Larry Wertz.

And so was my marriage, just like that.

Basically, I didn't keep my eye on what was important. I hadn't given enough attention to us. To maintaining us. To doing the things I needed to do to keep our union safe and strong. And because of this, I lost a beautiful woman, a beautiful friendship, and a life I thought was beautiful too.

This is a risky analogy. Still, it's one I see as indisputable. If we don't monitor our resources and assess their state of health enough, and if we don't keep our eye on the eco-ball and do what needs to be done, and if we don't protect what's going on and give wildness the attention it deserves, we risk falling into an indifferent slump, then getting burned on the most important thing we've got: a system that's fragile as hell—which can be destroyed in a nanosecond.

All it takes is an oil spill here, a hurricane there, or random fracking fluids spilling into ditches—like the 5500 gallons of diesel-based drilling mud that glooped its way into Cadron Creek when a truck overturned later that summer. Those toxins then made their way to the Arkansas

River, leaving an invisible trail of consequences in their wake. I write "invisible" because nobody knows what they did, and nobody seems to care. The reports of that disaster were just as ephemeral as the impression they made on the communities ignoring them. Both the spill and the minimal news of it were over in a flash and we got back to business as usual. Which, for me, meant experiencing my own manmade disaster: weeks of crying, shaking, panic attacks, and trying to make sense of sixteen years just flushed away like human waste.

Yes, this is an emotional argument. I'm connecting my own misfortune with what we have to lose—because right now, I'm a loser. But I don't mean that in a melodramatic sense; I mean it in a realistic sense. I'm feeling the loss that results from neglect and lack of foresight, which can affect fish as much as humans. Because the true bottom line is this: if you choose to ignore the obvious signals of a system in distress—when pipelines can burst at any instant—you can end up crushed and alone on a toxic lake.

And let me tell you: that's a pathetic place to be.

* * *

But like I said, it takes maintenance to keep something strong, healthy, able to endure. It takes a proactive will. You have to monitor what you value, and if you don't respect that enough to take preemptive action, you've essentially got yourself to blame. Because how can you blame others if you're not doing jack?

In Nicaragua, they're not doing jack. That's what I saw. The fishermen there are taking as much gar as they can get, despite their size, despite the season, despite all the alarms going off. My fear is that this exotic and diverse network of interconnected organisms is sliding down a slippery slope due to a lack of restrictions. As Paul Greenberg writes in *Four Fish*, "There is a general rule when it comes to overfishing. If no regulations are put in place, the more fish populations decline, and the more extreme and ecologically damaging fishing methods get."

Nicaragua, therefore, needs to step up its game by taking an example from Canada, where restrictions have worked to rebuild fisheries. Take the Grand Banks cod stocks, for example, in which two billion cod were harvested in the 1980s. By 1992, 95 percent of that population had been destroyed. The Canadian government then issued a complete closure of those fishing grounds, followed by decades of heavy restrictions. That population hasn't fully recovered yet, but CBS News reports a 69 percent growth since 2007.

A freshwater version of "ocean zoning" might be the answer. The idea is simple: Make certain places off limits to fishing so damaged populations can recover. This method has worked for salmon, it's worked for cod, and it's just as trendy as the concept of "ecosystem management," which, according to Greenberg again, "seeks to manage entire systems, modeling patterns for fishing and restoration that work toward reestablishing the balance of the many demands of prey and predator."

As for the rest of Central America, the consensus on tropical gar seems to be exactly what the Canadian Organization for Tropical Education and Rainforest Conservation claims it to be on the Caño Palma Biological Station website: those populations are threatened by "over-fishing, growing regional consumer demand and habitat alteration [that] has led to a decline in natural populations."

The vague status for alligator gar in Central America echoes this sentiment. There's not a lot of information on specific populations, but the fishermen and guides I spoke with in Costa Rica and Nicaragua say there are a lot fewer gator gar than there used to be. It's not known if alligator gar exist in El Salvador, Guatemala, and Honduras, but the general consensus is that they aren't there now. Hence, the only thing we know for sure about alligator gar in Central America is that we're lacking data.

In Mexico, it's also problematic for gar. According to a report in *Universidad y Ciencia* (vol. 18, no. 35), "fisheries have deteriorated significantly," mostly due to the recklessness of the oil industry, especially in the state of Tabasco. Aquaculture, however, has stepped in to supplement the demand for *pejelagarto,* and to preserve this fish not just as a symbol of cultural identity, but as a vital link in a collapsing chain that could lead to further crashes in the environment. Thus, the tropical gar is returning in Mexico as a farmed fish, but not so much in the wild.

As for alligator gar in Mexico, a June 2001 article in *The Southwestern Naturalist* notes that they range as far south as Veracruz, and that they are "highly valued as a food fish in northeastern Mexico." This article also notes that "Until now, no studies have been conducted on Mexican populations of the alligator gar," which, a dozen years later, still holds true. Roberto Mendoza, Carlos Aguilera, and Allyse Ferrara, however, authored a 2008 report on gar biology and culture in *Aquaculture Research* that explained how broodstock from the Aquaculture Center in Tampico, Tamaulipas, were used to repopulate nearby water bodies in the early 1980s. I asked Dr. Mendoza about the status of *el catan* over the last twenty years, and he replied that "the populations have increased a little bit after stocking alligator gar in different water bodies." This is encouraging news.

Regarding the Cuban gar, many sources claim it's an endangered species, but the International Union for Conservation of Nature and Natural Resources has not officially red-listed it as threatened in any way. The *Manjuari* is still limited to a very small range in Zapata Swamp and the Isla de la Juventud. It's a rare fish, which is why it's protected as a national treasure and pictured on a coin. Gar specialist Solomon David at the Shedd Aquarium in Chicago, who worked with this species at the University of Michigan, tells me that Andrés Hurtado and Tsai Garcia have been doing cutting-edge research on Cuban gar propagation and aquaculture, and their take-home message is that the *Manjuari*'s situation is dire due to swamp habitat being lost. Hence, the only place where the Cuban gar's numbers are quantifiably on the rise is in the aquarium trade.

In the United States, though, alligator gar are doing better than they have in a long time. New regulations in Arkansas, combined with increased research and attention to habitat and reintroduction have led to growth in populations. US Fish & Wildlife recently added seventy-eight young-of-year to the 2007 generation that is now supplementing the eighty-two big ones we have on record. Thanks, in part, to a concerted effort to bring their numbers up. When state and federal agencies join forces with universities, public and private laboratories, hatcheries, the media, fishing celebrities, and even freelance fish-writers, these are the results we can be proud of.

In Texas, new regulations and outreach have also had their effects, and the commercial fishing and bowhunting industries focused on gator gar have taken a big hit. Tons of alligator gar meat once destined for the meat grinder are not being harvested, and a lot fewer lunker gar are being left at the launches with holes through their heads. In 2014, due in part to a petition drive, a new law was passed that gives the Texas Parks & Wildlife Executive Director the authority to temporarily suspend fishing and hunting for gator gar in certain spawning areas.

Throughout the South, things are also looking up. As David Buck-meier, Nathan Smith, and Daniel Daugherty report in *Transactions of the American Fisheries Society* (no. 142), "Recently, all states in the Alligator Gar's historic U.S. range, with the exception of Louisiana, have either implemented harvest regulations or issued declarations of extirpation." Extirpation, of course, is no good—but that doesn't mean populations are going completely extinct. An optimistic way of looking at this is that when species begin to decline in an area, labeling them with classifications such as "imperiled," or issuing "declarations of extirpation" can help raise awareness and bring them back—like we did in Arkansas.

Fortunately, there's a lot of gar research going on. Gator gar life history, conservation, management, and age and growth are being studied at Auburn University in Alabama. Genetics, spawning, intensive culture, production techniques, sperm preservation, and tagging are being researched at Warm Springs National Fish Hatchery in Georgia. In Florida, the Fish and Wildlife Research Institute is investigating alligator gar habitat use, movement patterns, and distribution in the Escambia River. In Louisiana, LSU has teamed up with US Fish & Wildlife for gar research, and Nicholls State University is still going strong with status assessments, life histories, reproduction, and aging and growth investigations. The Mississippi Department of Fish and Wildlife is doing status assessments and surveys, spawning studies, telemetry research, and habitat assessment at the Private John Allen National Fish Hatchery in Tupelo. The University of Southern Mississippi is studying population genetics, tracking, and physiology. Oklahoma State University is looking at distribution, abundance, movements, habitat, and population dynamics for alligator gar in the Arkansas and Red River, and the Tennessee Wildlife Resources Agency is involved in gator gar management and restoration. Alligator gar are being stocked in Alabama, Georgia, Mississippi, Tennessee, Arkansas, and other southern states as well.

There's also a lot of gator gar stocking in the Midwest. The Mingo National Wildlife Refuge in Missouri can definitely be declared a successful alligator gar sanctuary. In 2013, the first gator gar was caught in Illinois since 1966. Following that, the Illinois Department of Natural Resources released 650 fingerlings into Crane Lake and the Little Sangamon River. Alligator gar have also been stocked as far north as Spunky Bottoms, Illinois, which is on the same latitude as the southern border of Iowa. According to the Indiana Division of Fish and Wildlife, "Because they have been stocked into the Ohio River, there is a possibility that alligator gar are either already in Indiana or will be found here in the future."

As the Nature Conservancy noted in a March 2013 article entitled "Big Fish: Return of the Alligator Gar," "Research is a key component to the reintroduction." Matt Miller, author of the article, remarks, "conservationists, anglers and naturalists have found a new and growing appreciation for the fish: a gar renaissance, if you will."

That's what I see as well. The granddaddy of the gars is making a comeback in the United States.

In the Third World, though, most populations seem to be going down, so it's not a totally rosy story. There's still a lot of work to do, and there's still a lot of fascination to take place if we want this part of our past to be part of our future.

* * *

I haven't caught an Arkansas alligator gar on rod and reel yet, but I've caught them in nets from one inch long to seven-foot-one. I've assisted others reeling in gator gar, and I've leapt into rivers to scoop them up with my bare hands. I caught a six-and-a-half footer on the Trinity, a three-and-a-half footer in Thailand, and a tropical on the border of Costa Rica. I've caught shortnose, spotteds, longnose, and a Florida gar in the Everglades.

But I also caught catfish along the way, especially one so huge I couldn't even weigh it, because my heavy-duty scale was broken. I caught that jet-black fatty on my trotline right before the pipeline burst.

Photo 61. Sixty-Pound Flathead.

Photo by Robin Becker.

I caught that catfish on a big Daiichi circle hook baited with a bream, and I caught plenty other cats ranging from twenty-one to thirty-eight pounds with hooks supplied by Daiichi. They gave me that tackle for gar, but admittedly, I haven't caught an alligator gar on any of their hooks yet. But I have caught longnose.

Like the one I landed at the end of the summer, out on a sandbar in the Arkansas River, using a six-inch channel cat as bait. When that gar hit, I sprang into action, releasing the drag on my Ambassadeur baitcaster at just the right moment. It took out a hundred yards, sometimes speeding up, sometimes slowing down—which is always a good sign. Then it stopped, so I did too, by taking a seat and pouring myself a gin and tonic.

Scotty laughed at my cockiness—that I'd actually take a break when I had a gar on the line. But as I've learned from fishing this river, it pays to wait 'em out. So five minutes later, when the spool started turning again, I put down my drink, got up, and set that hook, which doesn't always work when a fish is a football field away. So I ran with my rod until the slack ran out and the rod arced, meaning there was weight on the other end. Weight that came whooshing through the water with little resistance as gar tend to do, because their torpedo-shaped shells are aquadynamic.

It was the one out of ten, the gar that connects, according to the hit-and-miss ratio I've come to expect. And when I finally horsed it up on shore—a healthy, chromy, four-and-a-half-foot longnose hooked right through the bill—it was a bonus. Because these days, my only expectation is to enjoy the company I'm fishing with. So to finish the summer with that prehistoric fish grinning at me, that was my reward for studying gar, fishing for gar, flying all over the world for gar. At least, that's how I felt in that instant, and even if that's just me romanticizing gar, I'll take it.

I'll also take the arapaimas I caught in Thailand. And the tarpon I hooked in Central America and Mexico, which got away. And that gator I snagged in Florida, and that rattlesnake I saw in Texas. Because fishing for gar isn't just about fishing for gar; it's also about meeting other creatures along the way—which is a privilege, an honor, an act of love.

My only regret is not being able to close the deal with one of those heavy-duty rod and reel combos sent to me by Penn. I especially like the 330-size baitcaster, which I rigged with hundred-pound woven test. It's smooth and huge and powerful enough to crank in an elephant, and I'm still fishing with it. It's equipment I can get behind. In fact, I've been behind that reel nine times with alligator gar on the other end—so the next time, that'll be the one out of ten.

What worries me, though, are the bowfishermen, especially those on the stretch of river where I let my hybrid go. Every night I'm out on that sandbar, they're out there patrolling the opposite shore with high-intensity lights. Though the odds that they'll spot my hybrid are equivalent to finding a needle in a haystack, they see gator gar frequently and I wonder if they can restrain themselves. Are these the "weekend warriors" that the local tournament shortnose shooters claim don't care what they kill? Or are they just going for buffalo or carp? Or are these the guys in that photograph at Bates Baits in Mayflower, gloating over a full-sized alligator gar flopping off both ends of a tailgate? That one was shot in Lollie Bottoms, where three US Fish & Wildlife transmitters were found discarded in a ditch.

And what of Lollie Bottoms, where a giant pipe has been laid in a trench gouged from the earth this summer? It's been two years of sewage system construction now, and I'm sure that the gator gar that used to spawn in those fields have given up on Tupelo Bayou. With all that turbidity coming down, and with the droughts we've had in the last three years, it's highly unlikely that the right water temperature converging with the right water level and the right amount of flow could've enticed the big ones up during the spring floods. Since part of the Pool 7 population includes the largest known population of alligator gar in the state, I wonder where they go now that those spawning grounds have been rendered unusable.

I've been searching the backwaters of Bigelow Island, where a few years back I saw some alligator gar spawn stranded in puddles, waiting for the river to rise. Since that summer, I've seen longnose and shortnose fingerlings in that area, but I haven't seen any young gator gar. I check every summer at least a few times, when the water's high enough to get through the jetty where there are always large gar rolling in a place I like to swim. But in the last few years, I haven't seen any alligator gar there, where they used to be without fail.

Hence the questions: Where do they go? What can we do? But most of all—in a world where the cliché "there's no silver bullet" is commonplace on the network news—what's the solution?

* * *

Being the rebel I've always been, I hate to say it, but the solution, I think, is to maintain a strong economy. Capitalism is here to stay, and there's no changing the mainstream. That's the system we're working with, and it's part of our ecosystem as well. I might be naïve, but it seems to me that irrespective of which political party is in power, the stronger our governments are, the more they trickle down. Stronger economies mean more tax dollars, which translates into stronger schools and stronger communities, which results in more awareness of what's real. Stronger economies also mean more funding for science and research, which makes its way to fish. And as we're seeing with gar, the more we study them, the more good we do for them, and by extension, the more good we do for ourselves.

Stronger economies also mean stronger ideas about how to adapt to our changing environment, and consequently, stronger ideas about how to make the shift to alternative energy sources—or, at least, to less-damaging methods, which consider the long-term effects of what we do now. Because what we're doing now with pipelines and oil rigs and tankers and burning mass coal and mountain-top removal and strip-mining and hydrofracking, just for starters, is jeopardizing millions of species.

Perhaps I'm veering into the emotional argument again rather than a fact-based rationale, but guess what? That's where we're at in this book. We're at the end. All the authorities have been quoted. The statistics are over. No more metaphors. No more examples, no more anecdotes, no more rhetoric about how we have the power to make the common-sense changes we need to survive.

The moral of this story is simple: The more we know, the less stupid we are. And the less stupid we are, the less likely we are to keep repeating stupid mistakes. That's just the way it is, and everyone knows it.

If you ask me, our best defense against losing what's most important, including our identities and our existence, is to keep feeding the imagination by preserving the most fantastic elements in our midst. Like the alligator gar, a creature so strange and provocative of nightmares and dreams that it makes us ask who our monsters really are.

If we look at gar close enough, we see ourselves in the mirror. And if we look in the mirror close enough, we begin to see where we fall short. The trick, of course, is to put these reflections into action. The trick is to actually accept responsibility, and literally, physically, care for something. The trick is to change—our attitudes, our approaches, our priorities—or else we stagnate and lose momentum.

But really, there is no moral to any story. There's just the fact that when we concentrate on a problem, we can fix it. Like we're doing in Arkansas. Like we're doing throughout the American South and even in the Midwest—for alligator gar, a fish that's visible, tangible, fishable, and exceptional for providing balance.

The question for me, however, is whether we have the will to change ourselves in places beyond our borders—for a fish that hasn't changed in over a hundred million years. Because what happened in the United States is happening now in Mexico and Central America, where alligator and tropical gar are paying for our negligence.

Still, gar have adapted; it's in their nature. If we can do the same, I'm not so sure. The proof will be in our politics and how we handle a crisis we're afraid to fully envision, since enough of us can't conceive of how we relate, and what we owe, to what we call "the great outdoors." As if all you have to do to visit nature is turn a knob and then you're there. As if the environment knows how to take care of itself when we're meddling with it on a daily basis.

It could very well be that our rising level of carbon dioxide is increasing in direct proportion to our declining level of self-respect. It could very well be that we are on the cusp, that we are on the edge, and that what happens in the short-term will define what happens after that in ways that could cause a Lake Nicaragua of regret. Or just as bad, a Lake Conway of regret. Or even worse, a State of Tabasco state of regret.

For the moment, though, it's fair to say that longnose, shortnose, spotted, and Florida gar are doing fine. And in the United States, alligator gar have returned to some of their old hunting grounds.

We're learning from our mistakes. We're applying patches where we can. And even though we screw up, we are idealistic enough to believe that we have the power to change the world. And we can.

In the meantime, I'm going fishing—if only to see the gar roll.

Garpendix

The following schedule for the 2010 International Gar Conference is provided to show an overview of the cutting-edge gar research currently going on in the global gar community, and to credit some of the lead researchers in *Lepisosteid* research directly with their work. This information was taken from the URL www.nicholls.edu/bayousphere/ workinggroup/Program.pdf. Stay tuned to this website for information on the next international gar conference. Printed with permission from the International Network for Lepisosteid Fish Research and Management.

PROGRAM AGENDA FOR THE JOINT MEETING OF THE INTERNATIONAL NETWORK FOR LEPISOSTEID FISH RESEARCH AND MANAGEMENT

AND THE SOUTHERN DIVISION OF THE AMERICAN FISHERIES SOCIETY ALLIGATOR GAR TECHNICAL COMMITTEE

MAY 25–28, 2010

NICHOLLS STATE UNIVERSITY
THIBODAUX, LOUISIANA
TUESDAY MAY 25, 2010: Room 201 Gouaux Hall, Nicholls State University (NSU)

10:00-4:00 Gar Aging Workshop (Lunch Provided; Pre-registration required)

The goals of this workshop are to compare current techniques for aging garfish, to develop or recommend standard techniques and to identify future research needs.

Workshop participants are asked, but are not required, to bring hard structures for aging, to include but not limited to otoliths, otolith sections, branchiostegal rays, scales, scale sections, fin spines, fin spine sections, and digital images of hard structures. Isomet saws and microscopes will be available for use by participants. If participants need special equipment to demonstrate methods, please contact Allyse Ferrara.

WEDNESDAY MAY 26, 2010: 101 Gouaux Hall, NSU

8:30-4:00 Registration Open

9:30-10:00 Welcome and Introductory Remarks. Allyse Ferrara and Dr. Stephen Hulbert, President of Nicholls State University

10:00-10:50 Key Note Address. Season of the Gar: Exploring the Ecotone between Science, History, Lepisosteid Management, and Creative Nonfiction. Mark Spitzer

10:50-11:20 Restoration of Garfish *Atractosteus tropicus* (Pisces: Lepisosteidae) in the Refugio Nacional de Vida Silvestre Caño Negro, Costa Rica: A New Alternative for its Management and Conservation in Costa Rica. M. Protti Q., G. Márquez–Couturier, A. Sevilla C., and J.B. Ulloa R.

11:20-11:40 BREAK

11:40-12:00 Current Alligator Gar Management Activities in Alabama. David L. Armstrong and Ryan Peaslee

12:00-12:20 Pre- and Post-Regulation Harvest Rates of Alligator Gar *Atractosteus spatula* at Trinity River Bowfishing Tournaments. Dan Bennett

12:20-1:50 LUNCH

1:50-2:10 Movements and Habitat Use of Adult Alligator Gar in a Tributary of the Arkansas River. Edward R. Kluender, Lindsey Lewis, and Reid Adams

2:10-2:30 Movement and Habitat Use of Alligator Gar in the Trinity River, TX. Nathan G. Smith, David L. Buckmeier, and Daniel J. Daugherty

2:30-2:50 A Preliminary Analysis of Range-Wide Population Structure in Alligator Gar. Gregory R. Moyer and Brian R. Kreiser

2:50-3:20 Influence of Anabolic Hormones on Alligator Gar Breeders and Their Effect in Larvae. Roberto Mendoza and Carlos Aguilera

3:20-3:40 BREAK

3:40-5:40 Alligator Gar Technical Committee Meeting (Everyone welcome to attend)

THURSDAY MAY 27, 2010: 101 Gouaux Hall, NSU

7:30-4:00 Registration Open

8:00-8:10 Welcome and Introductory Remarks. Allyse Ferrara

8:10-8:30 Effect of Grading Frequency on Production of Alligator Gar Fingerlings in Tanks. Steve E. Lochmann and Lael A. Will

8:30-8:50 Evaluation of Structure, Forage and Stocking Density on Fingerling Production of Alligator Gar *Atractosteus spatula*. Peter Perschbacher

8:50-9:30 Characterization of the Supply Network of the Tropical Gar (*Atractosteus tropicus*) in Tabasco, Mexico. Vázquez-Navarrete, C.J. and G. Márquez-Couturier

9:30-9:50 Optimal Feed Rates for Juvenile Alligator Gar *Atractosteus spatula* Reared in Recirculating Systems. Tim A. Clay, Mark D. Suchy, Wendell Lorio, Allyse M. Ferrara, and Quenton C. Fontenot

9:50-10:10 BREAK

10:10-10:50 Preliminary Results of the *Atractosteus tropicus* (Pisces: Lepisosteidae) Larvae Rearing Using Two Different Culture Systems in Costa Rica. M. Protti Q., G. Márquez-Couturier, A. Sevilla C., and J.B. Ulloa R.

10:50-11:10 Preliminary Results of a Florida Gar *Lepisosteus platrhynchus* Age and Growth Study. Gintas Zavadzkas, Tim Clay, Quenton Fontenot, and Allyse Ferrara

11:10-11:30 Countergradient Variation in Growth of Spotted Gar (*Lepisosteus oculatus*) from Different Latitudes, with Implications for Conservation. Solomon David, R. Kik IV, M.J. Wiley, E.S. Rutherford, and J.S. Diana

11:30-11:50 Effects of Ambient Salinity on Plasma Osmolality of Juvenile Alligator Gar *Atractosteus spatula*, Spotted Gar *Lepisosteus oculatus*, Paddlefish *Polyodon spathula*, and Lake Sturgeon *Acipenser fulvescens*. Quenton C. Fontenot, Mark Suchy, Tim Clay, Ricky Campbell, Wendell Lorio, and Allyse Ferrara

11:50-12:10 Effects of Salinity Acclimation on Growth, Plasma Osmolality, and Metabolic Rate of Juvenile Alligator Gar. Daniel E. Schwarz and Peter J. Allen

12:10-1:30 LUNCH SERVED ON SITE (Plantation Suite of Student Union)

1:30-1:50 Standard Metabolic Rate of Alligator Gar *Atractosteus spatula* at Three Temperatures. Nick Barkowski, Brandon Baker, Brett Timmons, Alf Haukenes, and Steve E. Lochmann

1:50-2:10 Effects of Salinity on Growth and Survival on Larval and Juvenile Alligator Gar *Atractosteus spatula*. Quenton C. Fontenot, Mark Suchy, Tim Clay, Wendell Lorio, and Allyse Ferrara

2:10-2:30 Mercury Concentrations in the Muscle Tissue of Longnose Gar *Lepisosteus osseus* in Coastal North Carolina with Additional Contributions to the Life History. Jillian H. Osborne and R.A. Rulifson

3:10-3:40 Functional Analysis of *sox9* in the Spotted Gar *Lepisosteus oculatus*. Angel Amores, John Postlethwait, Yi-Lin Yan, Julian Catchen, Allyse Ferrara, and Quenton Fontenot

3:40-3:50 BREAK

3:50-4:10 Use of Digestive Physiology to Design of Microdiets for the Larviculture of Tropical Gar *Atractosteus tropicus*. C.A. Frías-Quintana, C.A. Álvarez-González, N. Perales-García, G. Márquez-Couturier, and W.M. Contreras-Sánchez

4:10-4:30 Tropical Gar, *Atractosteus tropicus*, Culture in Southeastern Mexico. W.M. Contreras-Sánchez, G. Márquez-Couturier, U. Hernández-Vidal, A. Hernández-Franyutti, C.A. Alvarez-Gonzalez, S. Paramo-Delgadillo, and L. Arias-Rodríguez

4:30-5:00 Physiological Response of Alligator Gar (*Atractosteus spatula*) to Pollution. Carlos Aguilera, Julio Cruz, Roberto Mendoza, and Ramón Chacón

6:00-9:00 Crawfish Boil at the Nicholls State University Farm with Live Cajun Music Performed by Treater

FRIDAY MAY 28, 2010: 101 Gouaux Hall, NSU

8:30-10:00 Registration Open

9:00-9:20 Bacteriocidal Activity of Spotted Gar Serum Mediated by Complement Protein. Justin Merrifield and Rajkumar Nathanial

9:20-9:40 Reproductive Characterization of Spotted Gar *Lepisosteus oculatus* in the Upper Barataria Estuary, Louisiana. Olivia A. Smith, Allyse Ferrara, Quenton C. Fontenot, and Gary J. LaFleur, Jr.

9:40-10:00 Evaluating Habitat Utilization and Diet of the Threatened Spotted Gar (*Lepisosteus oculatus*) in Rondeau Bay with the Aid of Radiotelemtry and Gastric Lavage. William Glass, Lynda Corkum, and Nicholas E. Mandrak

10:00-10:20 Preliminary Analysis of Alligator Gar *Atractosteus spatula* and Spotted Gar *Lepisosteus oculatus* Diets Collected in a Drainage Canal in Port Sulphur, Louisiana. Rachel Ianni, Allyse Ferrara, and Quenton Fontenot

10:20-10:40 The Neurotoxic Potency of Gar Oocyte Extract Peaks at Spawning. Nicole Broussard, Hamilton Farris, Allyse Ferrara, and Gary LaFleur, Jr.

10:40-11:00 Strategies for the Commercial Pilot Scale Culture of Tropical Gar (*Atractosteus tropicus*) in Tabasco, Mexico. G. Márquez-Couturier, C.J. Vázquez-Navarrete, I.C. Olive-Alvarez, O. Olive-Alvarez, and C.A. Alvarez-González.

11:00-11:40 Small-Scale Experimental Culture and Cost Analysis of Tropical Gar *Atractosteus tropicus* in Earthen Ponds in Tabasco, Mexico. Ulises Hernández-Vidal, Alejandro Macdonal-Vera, Juan M. Vidal-López, Wilfrido M. Contreras-Sánchez, and Arlette A. Hernández-Franyutti

11:40-12:00 Tropical Gar *Atractosteus tropicus* Culture in PVC-Lined Circular Tanks in Tabasco, Mexico. Ulises Hernandez-Vidal, Alejandro Macdonal-Vera, Wilfrido M. Contreras-Sánchez, Otilio Mendez-Marin, Sergio Hernandez-Garcia, Lenin Arias-Rodriguez, and Arlette A. Hernandez-Franyutti

12:00-12:30 International Gar Network Business Meeting (All are welcome to attend)

POSTER SESSION: 101 Gouaux Hall Lobby, NSU

Early Growth and Survival of Larval Alligator Gar *Atractosteus spatula* Reared on Artificial Floating Feed with or without a Live *Artemia* spp. Supplement. Tim A. Clay, Mark D. Suchy, Wendell Lorio, Allyse Ferrara, and Quenton C. Fontenot

Allometric Growth in Cuban Gar *Atractosteus tristoechus* Larvae. Yamilé Comabella, Julia Azanza, Andrés Hurtado, and Tsai García-Galano

Molecular Evolution of the Inhibin α-Subunit in Holostean Fishes. G.L. deGravelle, B.C. Moore, M.I. McClellan, and J.A. McLachlan

Alternate Aging Techniques for Alligator Gar. Kayla DiBennedetto

Comparison of Non-Linear Modeling for Alligator Gar Growth. Lin Xie, Peter Perschbacher, and Steve Lochmann

Alligator Gar *Atractosteus spatula* Intensive Culture Program at Warm Springs National Fish Hatchery, Warm Springs, Georgia. Jaclyn Zelko and Carlos Echevarría.